Discrete Dynamical Systems

Oded Galor

Discrete Dynamical Systems

 Springer

Acknowledgements

This book has evolved over the past decade while I have been teaching master and doctoral courses at Brown University and the Hebrew University of Jerusalem on the subjects of discrete dynamical systems and their applications for intertemporal macroeconomics and economic growth. I benefited from the interaction with several cohorts of students who have read the manuscript in its various stages and provided constructive comments on its different segments. I am especially grateful for Benny Berdugo, Ruben Durante, Leo Feler, Phillip Garner, Takuma Kunieda, Yishay Maoz, Petros Milionis and Carmina Vargas, who have carefully read and commented on the final draft of the manuscript.

Preface

This book provides an introduction to discrete dynamical systems – a framework of analysis that is commonly used in the fields of biology, demography, ecology, economics, engineering, finance, and physics.

The book characterizes the fundamental factors that govern the quantitative and qualitative trajectories of a variety of deterministic, discrete dynamical systems, providing solution methods for systems that can be solved analytically and methods of qualitative analysis for those systems that do not permit or necessitate an explicit solution.

The analysis focuses initially on the characterization of the factors that govern the evolution of state variables in the elementary context of one-dimensional, first-order, linear, autonomous systems. The fundamental insights about the forces that affect the evolution of these elementary systems are subsequently generalized, and the determinants of the trajectories of multi-dimensional, nonlinear, higher-order, non-autonomous dynamical systems are established.[1]

Chapter 1 focuses on the analysis of the evolution of state variables in one-dimensional, first-order, autonomous systems. It introduces a method of solution for these systems, and it characterizes the trajectory of a state variable, in relation to a steady-state equilibrium of the system, examining the local and global (asymptotic) stability of this steady-state equilibrium. The first part of the chapter characterizes the factors that determine the existence, uniqueness and stability of a steady-state equilibrium in the elementary context of one-dimensional, first-order, linear autonomous systems. Although linear dynamical systems do not govern the evolution of the majority of the observed dynamic phenomena, they serve as an important benchmark in the analysis of the qualitative properties of the nonlinear systems in the

---

[1] For continuous dynamical systems see Arnold (1973), Hirsch and Smale (1974), and Hale (1980).

proximity of steady-state equilibria. The second part of the chapter examines the trajectories of nonlinear systems based on the characterization of the linearized system in the proximity of a steady-state equilibrium. The basic propositions established in Chapter 1 provide the conceptual foundations for the analysis of multi-dimensional, higher-order, non-autonomous, dynamical systems.

Chapter 2 analyzes the evolution of a vector of interdependent state variables in multi-dimensional, first-order dynamical systems. It develops a method of solution for these systems, based on the construction of a time-independent transformation that converts the dynamical system into a new one that is characterized by either independent state variables whose evolution can be determined based on the analysis of the one-dimensional case, or partially dependent state variables whose evolution are determined by the well established properties of the Jordan matrix. The analysis of linear multi-dimensional dynamical systems provides an important reference point in the analysis of multi-dimensional nonlinear systems in the proximity of their steady-state equilibrium. It provides the characterization of the linear approximation of multi-dimensional nonlinear systems around steady-state equilibria.

Chapter 3 characterizes the trajectory of a vector of state variables in multi-dimensional, first-order, linear dynamical systems. It examines the trajectories of these systems when the matrix of coefficients has real eigenvalues and the vector of state variables converges or diverges in a monotonic or oscillatory fashion towards or away from a steady-state equilibrium that is characterized by either a saddle point or a stable or unstable (improper) node. In addition, it examines the trajectories of these linear dynamical systems when the matrix of coefficients has complex eigenvalues and the system is therefore characterized by a spiral sink, a spiral source, or a periodic orbit.

Chapter 4 analyzes the trajectory of a vector of state variables in multi-dimensional, first-order, nonlinear systems. It utilizes the characterization of linear multi-dimensional systems to examine the trajectory of the nonlinear systems in light of the *Stable Manifold Theorem*. In particular, the analysis examines the properties of the local stable and unstable manifolds, and the corresponding global stable and unstable manifolds.

Chapter 5 characterizes the evolution of a vector of state variables in higher-order as well as non-autonomous systems. It establishes the solution method for these higher-order and non-autonomous systems and it analyzes the factors that determine the qualitative properties of these discrete dynamical systems in the linear and subsequently the nonlinear case. The analysis is based upon the transformation of higher-order and non-autonomous systems into a multi-dimensional first-order systems that can be examined based on the analysis in Chaps. 2–4. In particular, a one-dimensional second-order system is converted into a two-dimensional first-order system, a one-dimensional third-order system is transformed into a three-dimensional first-order system, a one-dimensional n^{th}-order system is converted into an n-dimensional first-order system, and an n- dimensional m^{th}-order system is transformed into an $n \times m$- dimensional first-order system. Similarly, the analysis of non-autonomous systems is based on their transformation into higher-dimension, time-independent (autonomous) systems that can be examined based on the analysis of multi-dimensional, first-order systems in Chaps. 2–4.

Chapter 6 provides a complete characterization of several representative examples of two-dimensional dynamical systems. These examples include a first-order linear system with real eigenvalues, a first-order linear system with complex eigenvalues that exhibits a periodic orbit, a first-order linear system with complex eigenvalues that exhibits a spiral sink, a first-order nonlinear system that is characterized by a oscillatory convergence, and a second-order one-dimensional system converted into a first-order, two-dimensional system characterized by a continuum of equilibria and oscillatory divergence.

The book is designed for advanced undergraduate and graduate students in the fields of demography, ecology, economics, engineering, evolutionary biology, finance, mathematics, and physics, who are familiar with differential calculus and linear algebra. Furthermore, it is a useful reference for researchers of applied disciplines in which discrete dynamical systems are commonly employed.

Providence, USA
September 2006   Oded Galor

Contents

1

One-Dimensional, First-Order Systems

This chapter analyzes the evolution of a state variable in one-dimensional, first-order, discrete dynamical systems. It introduces a method of solution for these systems, and it characterizes the trajectory of the state variable, in relation to its steady-state equilibrium, examining the local and global (asymptotic) stability of this steady-state equilibrium.

The first part of the chapter characterizes the factors determining the existence, uniqueness and stability of a steady-state equilibrium in the elementary context of one-dimensional, first-order, linear autonomous systems. Although linear dynamical systems do not necessarily govern the evolution of the majority of the observed dynamic phenomena, they serve as an important benchmark in the analysis of the qualitative properties of nonlinear systems, providing the characterization of the linear approximation of nonlinear systems in the proximity of steady-state equilibria. The second part of the chapter examines the trajectories of nonlinear systems based on the characterization of the linearized system in the proximity of a steady-state equilibrium.

The basic propositions derived in this chapter provide the conceptual foundations for the generalization of the analysis and the characterization of multi-dimensional, higher-order, non-autonomous, dynamical systems.

The qualitative analysis of these dynamical systems is based upon the examination of the factors that determine the actual trajectory of the state variable. However, as will become apparent, once the basic propositions that characterize the properties of these systems are derived, an explicit solution is no longer required in order to characterize the nature of these dynamical systems.

1.1 Linear Systems

Consider a one-dimensional, first-order, autonomous, linear difference equation that governs the evolution of a state variable, y_t, over time.

$$y_{t+1} = ay_t + b, \qquad t = 0, 1, 2, 3, \cdots, \tag{1.1}$$

where the value of the *state variable* at time t, y_t, is a real number, i.e., $y_t \in \Re$, the parameters a and b are constant real numbers, namely $a, b \in \Re$, and the initial value of the state variable at time 0, y_0, is given.[1]

The system is defined as a one-dimensional, first-order, autonomous, linear difference equation since it describes the evolution of a *one-dimensional* state variable, y_{t+1}, whose value depends in a *linear* and *time-independent* (autonomous) fashion on its value in the pervious period (first-order), y_t.

1.1.1 Characterization of the Solution

A solution to the difference equation $y_{t+1} = ay_t + b$ is a *trajectory* (or an *orbit*) of the state variable, $\{y_t\}_{t=0}^{\infty}$, that satisfies this law of motion at any point in time. It relates the value of the state variable at time t, y_t, to its initial value, y_0, and to the parameters a and b.

The derivation of a solution may follow several methods. In particular, the intuitive method of iterations generates a pattern that can be easily generalized to a solution rule.

Given the value of the state variable at time 0, y_0, the dynamical system $y_{t+1} = ay_t + b$ implies that the value of the state variable at time 1, y_1, is

$$y_1 = ay_0 + b. \tag{1.2}$$

Given the value of the state variable at time 1, y_1, the value of the state variable at time 2, y_2, is uniquely determined.

$$y_2 = ay_1 + b = a(ay_0 + b) + b = a^2 y_0 + ab + b. \tag{1.3}$$

---

[1] Without loss of generality, the feasible domain of the time variable, t, is truncated to be the set of non-negative integers. Moreover, the initial condition is defined as the value of the state variable at time 0. In general, t can be defined to be an element of any subset of the set of integers, and the initial value of the state variable, y_0, can be given at any point within this interval.

Similarly, the value of the state variable at time $3, 4, ..., t$, is

$$y_3 = ay_2 + b = a(a^2 y_0 + ab + b) + b = a^3 y_0 + a^2 b + ab + b$$

$$\vdots \qquad\qquad \vdots \tag{1.4}$$

$$y_t = a^t y_0 + a^{t-1} b + a^{t-2} b + ... + ab + b.$$

Hence, for $t = 1, 2, ...,$

$$y_t = a^t y_0 + b \sum_{i=0}^{t-1} a^i. \tag{1.5}$$

Since $\sum_{i=0}^{t-1} a^i$ is the sum of the geometric series, $\{1, a, a^2, a^3, ..a^{t-1}\}$, whose factor is a, it follows that

$$\sum_{i=0}^{t-1} a^i = \begin{cases} \frac{1-a^t}{1-a} & if & a \neq 1 \\ t & if & a = 1, \end{cases} \tag{1.6}$$

and therefore

$$y_t = \begin{cases} a^t y_0 + b\frac{1-a^t}{1-a} & if & a \neq 1 \\ y_0 \quad + bt & if & a = 1. \end{cases} \tag{1.7}$$

Alternatively,

$$y_t = \begin{cases} [y_0 - \frac{b}{1-a}]a^t + \frac{b}{1-a} & if & a \neq 1 \\ y_0 + bt & if & a = 1. \end{cases} \tag{1.8}$$

Thus, as long as an initial condition of the state variable is given, the entire trajectory of the state variable is uniquely determined.

The trajectory derived in (1.8) reveals the qualitative role of the parameter a, and to a lesser extent, b, in the evolution of the state variable over time. These parameters determine whether the dynamical system evolves monotonically or in oscillations, and whether the state variable converges in the long run to a steady-state equilibrium, diverges asymptotically to plus or minus infinity, or displays a two-period cycle. Hence, a qualitative examination of a dynamical system requires the analysis of the asymptotic behavior of the system as time approaches infinity.

1.1.2 Existence of Steady-State Equilibria

Steady-state equilibria provide an essential reference point for a qualitative analysis of the behavior of dynamical systems. A *steady-state equilibrium* (alternatively defined as a *stationary equilibrium*, a *rest point*, an *equilibrium point*, or a *fixed point*) is a value of the state variable y_t that is invariant under the law of motion dictated by the dynamical system.

Definition 1.1. *(A Steady-State Equilibrium)*
A steady-state equilibrium of the difference equation $y_{t+1} = ay_t + b$ *is* $\bar{y} \in \Re$ *such that*

$$\bar{y} = a\bar{y} + b.$$

Thus, if the state variable is at a steady-state equilibrium, it will remain there in the absence of any perturbations of the dynamical system due to either changes in the parameters a and b or direct perturbations in the value of the state variable itself. Namely, if $y_t = \bar{y}$ then $y_s = \bar{y}$ for all $s > t$.

As follows from Definition 1.1, as long as $a \neq 1$, there exists a unique steady-state equilibrium $\bar{y} = b/(1-a)$ for the difference equation $y_{t+1} = ay_t + b$. However, given the linear structure of the dynamical system, if $a = 1$ and $b = 0$ then in every time t, $y_{t+1} = y_t$ and the state variable does not deviate from its initial condition. In particular, $y_t = y_{t-1} = y_{t-2} = \ldots = y_0$ and the system is in a steady-state equilibrium where $\bar{y} = y_0$. In contrast, if $a = 1$ and $b \neq 0$, a steady-state equilibrium does not exist and the state variable increases indefinitely if $b > 0$, or decreases indefinitely if $b < 0$.

Hence, following Definition 1.1,

$$\bar{y} = \begin{cases} \frac{b}{1-a} & if \quad a \neq 1 \\ \\ y_0 & if \quad a = 1 \text{ and } b = 0. \end{cases} \tag{1.9}$$

Thus, the necessary and sufficient conditions for the existence of a steady-state equilibrium are given by the values of the parameters a and b, as stated in (1.9), that permits the system to have a steady-state equilibrium.

Proposition 1.2. *(Existence of Steady-State Equilibrium)*
A steady-state equilibrium of the difference equation $y_{t+1} = ay_t + b$ *exists if and only if*

$$\{a \neq 1\} \ or \ \{a = 1 \ and \ b = 0\}.$$

Hence, given the steady-state level of the state variable, y_t, as derived in (1.9), the solution to the difference equation $y_{t+1} = ay_t + b$ can be expressed in terms of the deviations of the initial value of the state variable, y_0, from its steady-state value, $\bar{y}$. Namely, substituting the value of $\bar{y}$ into the solution given by (1.8), it follows that

$$y_t = \begin{cases} (y_0 - \bar{y})a^t + \bar{y} & if & a \neq 1 \\ \\ y_0 + bt & if & a = 1. \end{cases} \tag{1.10}$$

1.1.3 Uniqueness of Steady-State Equilibria

A steady-state equilibrium of the linear dynamical system, $y_{t+1} = ay_t + b$, is not necessarily unique. As depicted in Figs. 1.1, 1.3, 1.7, 1.9 and 1.10 for $a \neq 1$, the steady-state equilibrium is unique. However, as depicted in Fig. 1.5, for $a = 1$ and $b = 0$, a continuum of steady-state equilibria exists, reflecting the entire set of feasible initial conditions.

Necessary and sufficient conditions for the uniqueness of a steady-state equilibrium are given by the values of the parameters a and b, as stated in (1.9) that permits the system to have a distinct steady-state equilibrium.

Proposition 1.3. *(Uniqueness of Steady-State Equilibrium)*
A steady-state equilibrium of the difference equation $y_{t+1} = ay_t + b$ *is unique if and only if*

$$a \neq 1.$$

1.1.4 Stability of Steady-State Equilibria

The stability analysis of the system's steady-state equilibria determines whether a steady-state equilibrium is attractive or repulsive for all or at least some set of initial conditions. It facilitates the study of the local, and often the global, properties of a dynamical system, and it permits the analysis of the implications of small, and sometimes large, perturbations that occur once the system is in the vicinity of a steady-state equilibrium.

A steady-state equilibrium is *globally* (asymptotically) stable if the system converges to this steady-state equilibrium regardless of the level of the initial condition, whereas a steady-state equilibrium is *locally* (asymptotically) stable if there exists an ϵ- neighborhood of the steady-state equilibrium such that from every initial condition within this neighborhood the system converges to this steady-state equilibrium. Formally the definition of local and global stability are as follows:[2]

Definition 1.4. *(Local and Global Stability of a Steady-State Equilibrium)*
A steady-state equilibrium, $\overline{y}$, of the difference equation $y_{t+1} = ay_t + b$ is:

- *globally (asymptotically) stable, if*

$$\lim_{t \to \infty} y_t = \overline{y} \quad \forall y_0 \in \Re;$$

- *locally (asymptotically) stable, if*

$$\lim_{t \to \infty} y_t = \overline{y} \quad \forall y_0 \text{ such that } \ |y_0 - \overline{y}| < \epsilon \text{ for some } \epsilon > 0.$$

Alternatively, if the state variable is in a steady-state equilibrium and upon a sufficiently small perturbation it converges asymptotically back to this steady-state equilibrium, then this equilibrium is *locally* stable. However, if regardless of the magnitude of the perturbation the

---

[2] The economic literature, to a large extent, refers to the stability concepts in Definition 4.2 as global stability and local stability, respectively, whereas the mathematical literature refers to them as global asymptotic stability and local asymptotic stability, respectively. The concept of stability in the mathematical literature is reserved to situations in which trajectories that are initiated from an ϵ-neighborhood of a fixed point remain sufficiently close to this fixed point thereafter.

state variable converges asymptotically to this steady-state equilibrium, then the equilibrium is *globally* stable.

Global stability of a steady-state equilibrium necessitates the *global uniqueness* of the steady-state equilibrium. If there is more than one steady-state equilibrium, none of the equilibria can be globally stable since there exist at least two points in the relevant space from which there is no escape and convergence from each of these steady-state equilibria to the other steady-state equilibrium is therefore not feasible.

Proposition 1.5. *(Necessary Condition for Global Stability of Steady-State Equilibrium)*
A steady-state equilibrium of the difference equation $y_{t+1} = ay_t + b$ *is globally (asymptotically) stable only if the steady-state equilibrium is unique.*

Local stability of a steady-state equilibrium necessitates the *local uniqueness* of the steady-state equilibrium. Namely the absence of any additional point in the neighborhood of the steady-state from which there is no escape. If the system is characterized by a continuum of equilibria none of these steady-state equilibria is locally stable. There exists no neighborhood of a steady-state equilibrium that does not contain additional steady-state equilibria, and hence there exist initial conditions within an ε- neighborhood of a steady-state equilibrium that do not lead to this steady-state equilibrium in the long run. Thus, local stability of a steady-state equilibrium requires the local uniqueness of this steady-state equilibrium.

If the system is linear there is either unique steady-state equilibrium or continuum of (unstable) steady-state equilibria. Local uniqueness of a steady-state equilibrium therefore implies global uniqueness, and local stability therefore necessarily implies global stability.

As follows from the definitions of local and global stability, the stability of a steady-state equilibrium can be obtained by the examination of the properties of the system as time approaches infinity.

As follows from the solution for the difference equation $y_{t+1} = ay_t + b$, given by (1.10),

$$\lim_{t \to \infty} y_t = \begin{cases} [y_0 - \bar{y}] \lim_{t \to \infty} a^t + \bar{y} & \text{if} \quad a \neq 1 \\ \\ y_0 + b \lim_{t \to \infty} t & \text{if} \quad a = 1, \end{cases} \tag{1.11}$$

and therefore the limit of the absolute value of the state variable, $|y_t|$, is

$$\lim_{t \to \infty} |y_t| = \begin{cases} |\overline{y}| & if \quad \{|a| < 1\} \ or \\ & \qquad \{|a| > 1 \ \& \ y_0 = \overline{y}\} \\[2ex] |y_0| & if \quad a = 1 \ \& \ b = 0 \\[2ex] \left. \begin{matrix} |y_0| & for \quad t = 0, 2, 4, \dots \\[1ex] |b - y_0| & for \quad t = 1, 3, 5, \dots \end{matrix} \right\} & if \quad a = -1 \\[2ex] \infty & otherwise. \end{cases}$$

$$(1.12)$$

Thus, as follows from the property of the absolute value of the state variable y_t, as time approaches infinity, the absolute value of the parameter a and the value of b determines the long run value of the state variable. Moreover, the absolute value of the parameter a determines whether a steady-state equilibrium $\overline{y}$ is globally stable.

In particular, in the feasible range of the parameter a and b, the dynamical system exhibits five qualitatively different trajectories, characterized by the existence of a unique and globally stable steady-state equilibrium, a unique unstable, steady-state equilibrium, continuum of steady-state equilibria, inexistence of steady-state equilibria, and two-period cycles.

A. Unique Globally Stable Steady-State Equilibrium $(|a| < 1)$

If the coefficient $|a| < 1$, then the system is globally (asymptotically) stable converging to the steady-state equilibrium $\overline{y} = b/(1 - a)$, regardless of the initial condition, y_0. In particular, if $0 < a < 1$ then as depicted in the phase diagram in Fig. 1.1, the evolution of the state variable is characterized by monotonic convergence towards the steady-state equilibrium $\overline{y}$ regardless of the initial level of the state variable, y_0.

The steady state locus $y_{t+1} = y_t$ intersects with the linear difference equation, $y_{t+1} = ay_t + b$, at the steady-state equilibrium $\overline{y}$. Given y_0, the value of $y_1 = ay_0 + b$ can be read from corresponding value along the line $y_{t+1} = ay_t + b$. This value of y_1 can be mapped back to the y_t axis via the 45^0 line. Similarly, given y_1, the value of $y_2 = ay_1 + b$ can be read from the corresponding value along the line $y_{t+1} = ay_t + b$ and mapped back to the y_t axis via the 45^0 line. Hence, as depicted in Fig. 1.1, the state variable evolves along the depicted arrows of motion and converges monotonically to the steady-state equilibrium $\overline{y}$.

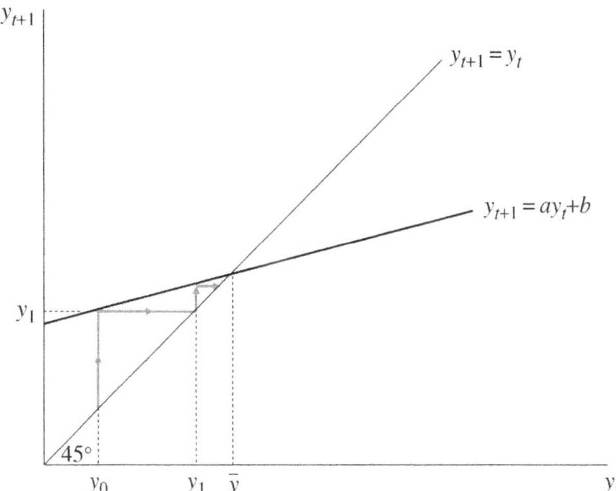

Fig. 1.1. One-Dimensional, First-Order, Linear System
Unique, Globally Stable, Steady-State Equilibrium
Monotonic Convergence
$0 < a < 1$

The state variable y_t converges monotonically from its initial level y_0 to the steady-state level $\bar{y}$. If $y_0 > \bar{y}$, then this monotonic convergence represents a declining value of y_t towards the steady-state level, y_0, If $y_0 < \bar{y}$, as depicted in Fig. 1.2 then this monotonic convergence represents an increasing value of y_t towards the steady-state level, $\bar{y}$.

If, however, the coefficient $-1 < a < 0$, then as depicted in Fig. 1.3 and 1.4, the convergence of the state variable to its steady-state value is oscillatory. The state variable, y_t converges in oscillations from its initial level y_0 to the steady-state level $\bar{y}$.

The oscillations of the state variable y_t subsides monotonically in the convergence process to the steady-state value $\bar{y}$, as depicted in Fig. 1.4.

B. Continuum of Unstable Steady-State Equilibria
$\{a = 1 \text{ and } b = 0\}$

If the coefficient $a = 1$ and the constant $b = 0$, the system, as depicted in Fig. 1.5, is characterized by a continuum of steady-state equilibria. Each steady-state equilibrium can be reached if and only if the system starts at this equilibrium and it is therefore neither globally nor locally stable. Any ε- neighborhood of a steady-state equilibrium contains other steady state equilibria and thus there exist initial conditions in any ε- neighborhood of a steady-state equilibrium that do not

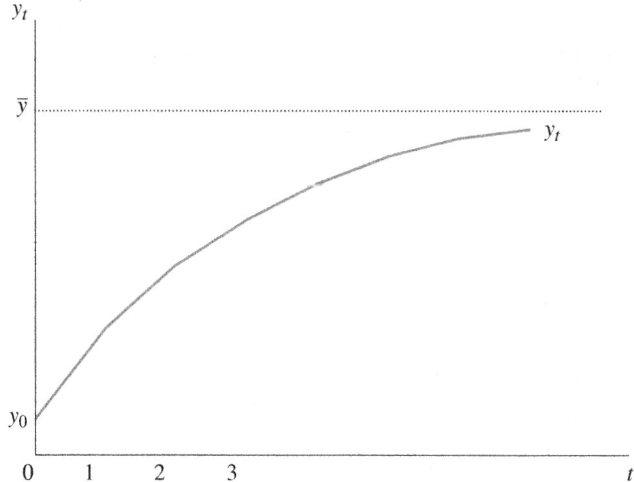

Fig. 1.2. The Evolution of the State Variable
Monotonic Convergence
$0 < a < 1$

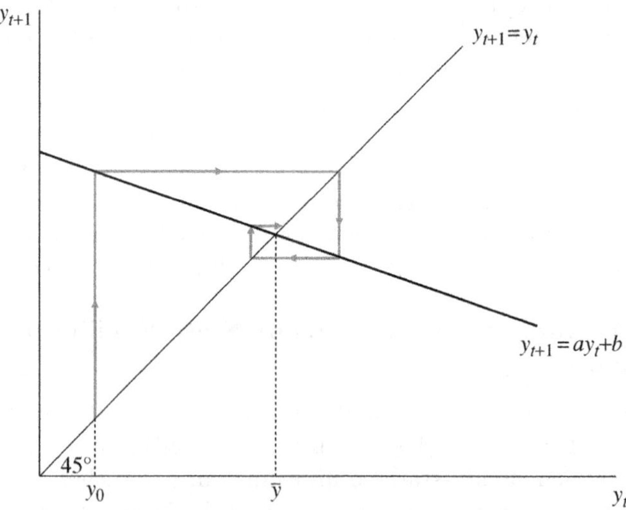

Fig. 1.3. One-Dimensional, First-Order, Linear System
Unique, Globally Stable, Steady-State Equilibrium
Oscillatory Convergence
$-1 < a < 0$

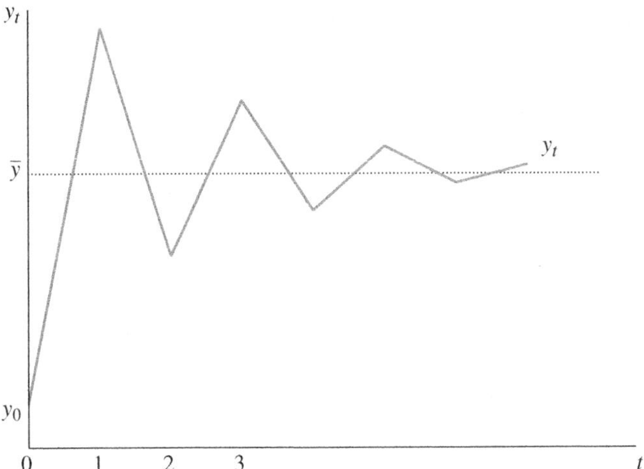

Fig. 1.4. The Evolution of the State Variable
Oscillatory Convergence
$-1 < a < 0$

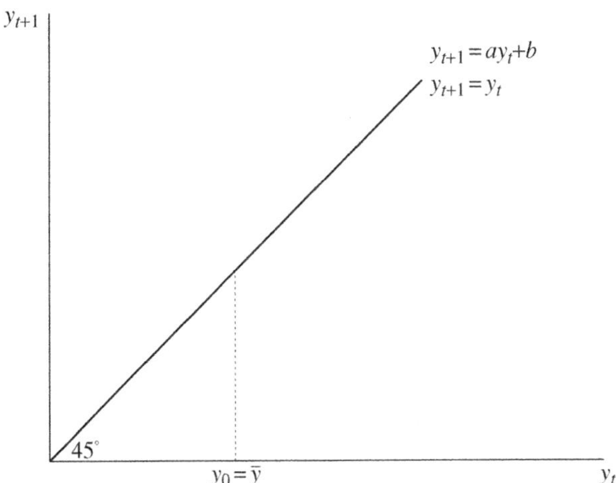

Fig. 1.5. One-Dimensional, First-Order, Linear System
Continuum of Unstable Steady-State Equilibria
$a = 1$ and $b = 0$

lead to this steady-state equilibrium in the long run. Each steady-state equilibrium is therefore unstable.

C. Non-Existence of a Steady-State Equilibrium
$\{a = 1 \text{ and } b \neq 0\}$

If the coefficient $a = 1$ and the constant $b \neq 0$, the system is characterized by the absence of a steady-state equilibrium and the state variable diverges to either plus or minus infinity. As shown in Fig. 1.6, if $b > 0$, then $\lim_{t \to \infty} y_t = +\infty$, i.e. the state variable increases monotonically over time and approaches asymptotically $+\infty$, whereas if $b < 0$, the value of the state variable declines monotonically over time and approaches asymptotically $-\infty$, i.e., $\lim_{t \to \infty} y_t = -\infty$.

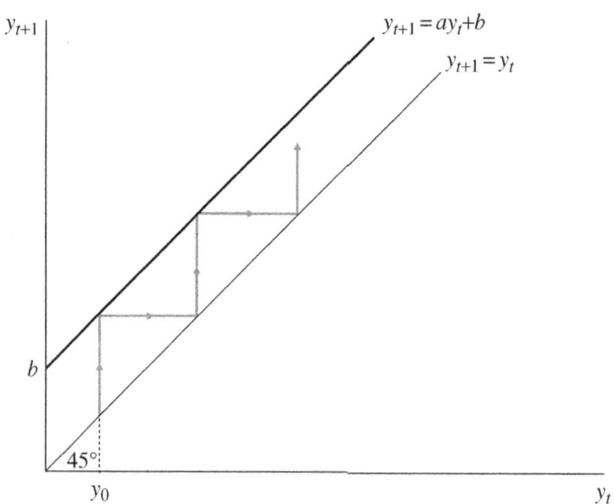

Fig. 1.6. One-Dimensional, First-Order, Linear System
Non-Existence of a Steady-State Equilibrium
$a = 1$ and $b \neq 0$

D. Two-Period Cycle $(a = -1)$

If the coefficient $a = -1$, then the system, as depicted in Fig. 1.7, is characterized by a continuum of (unstable) two-period cycles,[3] and

---

[3] Note that definition of stability is perfectly applicable for periodic orbits, provided that the dynamical system is redefined to be the nth iterate of the original one, where n is the periodicity of the cycle.

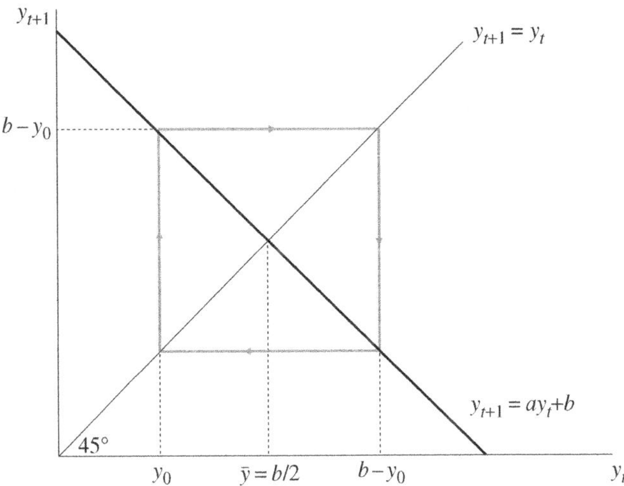

Fig. 1.7. One-Dimensional, First-Order, Linear System
Unstable Two-Period Cycle
$$a = -1$$

the unique steady-state equilibrium, $\bar{y} = b/2$, is unstable. The initial value of the state variable dictates the values of the two-period cycles.

Given the initial value of the state variable, y_0, then as depicted in Fig. 1.8, the state variable oscillates between y_0 and $b - y_0$ in a two-period cycle.

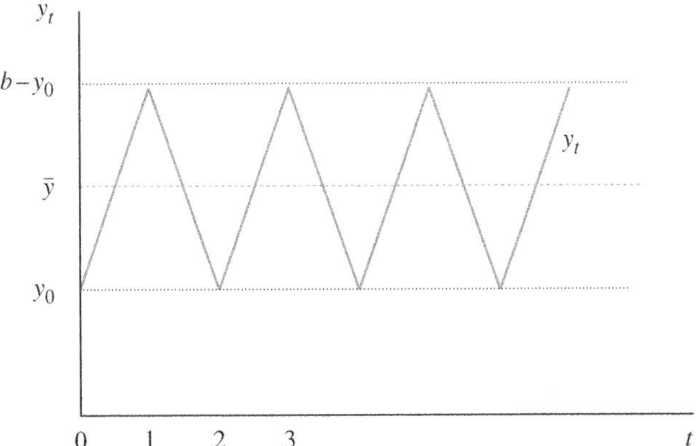

Fig. 1.8. The Evolution of the State Variable
Two-Period cycle
$$a = -1$$

There exists a continuum of two-period cycles that are determined by the level of the state variable in period 0. Each of these two-period cycles is unstable. Namely, unless the system originates in a given two-period cycle, the system will not reach it. Moreover, in a linear system, unless a two-period cycle exists at time 0, a two-period cycle will not exist.

E. Unique Unstable Steady-State Equilibrium ($|a| > 1$)

If the coefficient $|a| > 1$ then the system, as depicted in Figs. 1.9 and 1.10, is unstable. For $y_0 \neq b/(1-a)$, $\lim_{t\to\infty} |y_t| = \infty$, whereas for $y_0 = b/(1-a)$ the system starts at the steady-state equilibrium where it remains thereafter. Every minor perturbation, however, causes the system to step on a diverging path.

In particular, if the coefficient $a > 1$, then as depicted in Fig. 1.9, the state variable y_t diverges monotonically. If its initial condition is larger than $\bar{y}$, then it diverges monotonically to $+\infty$, whereas if the initial condition is smaller than $\bar{y}$, it diverges monotonically to $-\infty$.

However if the coefficient $a < -1$, then as depicted in Fig. 1.10, the state variable y_t diverges in oscillations. The oscillations of the state variable y_t are magnified monotonically in this process of divergence, oscillating in the limit between $+\infty$ and $-\infty$.

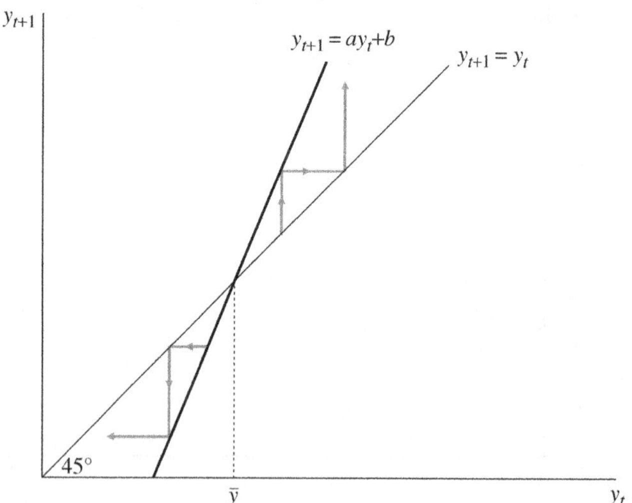

Fig. 1.9. One-Dimensional, First-Order, Linear System
Unstable Steady-State Equilibrium
Monotonic Divergence
$a > 1$

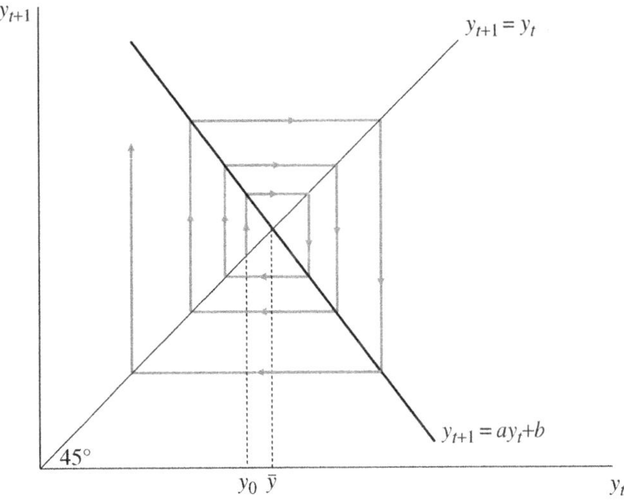

Fig. 1.10. One-Dimensional, First-Order, Linear System
Unstable Steady-State Equilibrium
Oscillatory Divergence
$a < -1$

The characterization of the asymptotic properties of the state variable y_t as $t \to \infty$, established in (1.12) and analyzed in Cases (A)–(E), provides the necessary and sufficient conditions for global stability of the linear system.

Proposition 1.6. *(A Necessary and Sufficient Condition for Global Stability of Steady-State Equilibrium)*
A steady-state equilibrium of the difference equation $y_{t+1} = ay_t + b$ *is globally stable if and only if*

$$|a| < 1.$$

Corollary 1.7. *(Necessary and Sufficient Conditions for Monotonic and Oscillatory Convergence)*
For any $y_0 \in \Re$,

$$\lim_{t \to \infty} y_t = \bar{y} \qquad if \quad |a| < 1,$$

where

$$\text{(a) convergence is monotonic if and only if}$$
$$a \in [0, 1);$$

$$\text{(b) convergence is oscillatory if and only if}$$
$$a \in (-1, 0).$$

1.2 Nonlinear Systems

This section analyzes the evolution of a state variable in a one-dimensional, first-order, nonlinear discrete dynamical system. It characterizes the evolution of the state variable in the proximity of a steady-state equilibrium based on a linear approximation of this nonlinear motion in the vicinity of a steady-state equilibrium. Subsequently, it provides some restrictive sufficient conditions for global stability of a nonlinear system.

Consider the one-dimensional autonomous, first-order, nonlinear difference equation that governs the evolution of a state variable, y_t, over time.

$$y_{t+1} = f(y_t), \qquad t = 0, 1, 2, \cdots, \tag{1.13}$$

where $f : \Re \to \Re$ is a continuously differentiable single-valued function and the initial value of the state variable, y_0, is given.[4]

1.2.1 The Solution

A solution to the difference equation $y_{t+1} = f(y_t)$, is a *trajectory* (or an *orbit*) of the state variable, $\{y_t\}_{t=0}^{\infty}$, that satisfies this law of motion at any point in time. It relates the value of the state variable at time t, y_t, to its initial value, y_0, and to the function f.

---

[4] For the local analysis, it is sufficient that the function $f : \Re \to \Re$ is continuously differentiable only in some neighborhoods of the relevant steady-state equilibrium.

Using the method of iterations, the trajectory of this nonlinear system, $\{y_t\}_{t=0}^{\infty}$, can be written as follows:

$$y_1 = f(y_0)$$

$$y_2 = f(y_1) = f[f(y_0)] \equiv f^{\{2\}}(y_0)$$

$$y_3 = f(y_2) = f[f(y_1)] = f[f^{\{2\}}(y_0)] \equiv f^{\{3\}}(y_0) \qquad (1.14)$$

$$\vdots \quad \vdots$$

$$y_t \equiv f^{\{t\}}(y_0),$$

where $f^{\{t\}}(y_0)$ is the outcome of t iterations of the function f over the initial condition y_0.

Unlike the solution to the linear system (1.1), the solution for the nonlinear system (1.14) is not very informative about the factors that determine its qualitative patterns and its tendency to converge to a steady-state equilibrium, to diverge to plus or minus infinity or to display a periodic orbit.

Hence, additional methods of analysis are required in order to gain an insight about the qualitative behavior of this nonlinear system. In particular, a linear approximation of the nonlinear system in the vicinity of a steady-state equilibrium is instrumental in the study of the qualitative behavior of nonlinear dynamical systems.

1.2.2 Existence, Uniqueness and Multiplicity of Steady-State Equilibria

Definition 1.8. *(A Steady-State Equilibrium)*
A steady-state equilibrium of the difference equation $y_{t+1} = f(y_t)$ *is a level* $\overline{y} \in \Re$ *such that*

$$\overline{y} = f(\overline{y}).$$

Generically, a nonlinear system may be characterized by the existence of unique steady-state equilibrium, the existence of multiplicity of (distinct) steady-state equilibria, the existence of chaotic behavior, or the non-existence of a steady-state equilibrium. Furthermore, the nonlinear system may converge to a steady-state equilibrium, may diverge to plus or minus infinity, may converge to a periodic orbit, and,

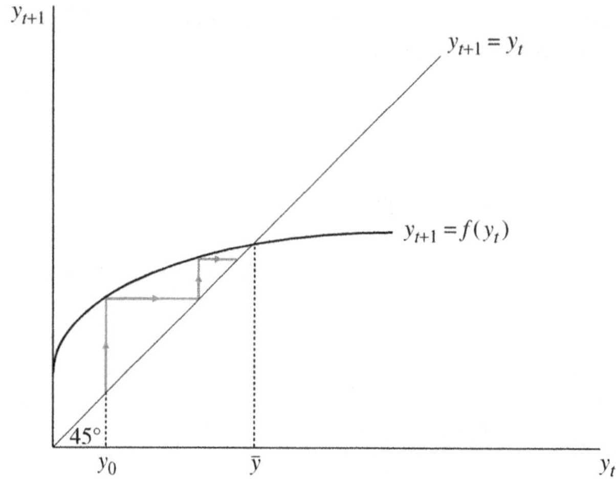

Fig. 1.11. One-Dimensional, First-Order, nonlinear System
Unique, Globally Stable, Steady-State Equilibrium: $\lim_{t\to\infty} y_t = \bar{y}$

unlike a linear system, a nonlinear system may exhibit chaotic behavior.[5] Figures 1.11–1.13 depict these various configurations under the assumption that $f : \Re_+ \to \Re$.

Figure 1.11 depicts a system with a globally stable unique steady-state equilibrium. Given y_0, the state variable evolves towards the steady-state equilibrium and $\lim_{t\to\infty} y_t = \bar{y}$.

Figure 1.12 depicts a system with multiple distinct steady-state equilibria: $\bar{y}_1$ and $\bar{y}_3$ are locally stable, whereas $\bar{y}_2$ is unstable. Namely,

$$\lim_{t\to\infty} y_t = \begin{cases} \bar{y}_1 & if & y_0 \in (0, \bar{y}_2) \\ \bar{y}_2 & if & y_0 = \bar{y}_2 \\ \bar{y}_3 & if & y_0 \in (\bar{y}_2, \infty). \end{cases} \tag{1.15}$$

Figure 1.13 depicts the dynamical system in the absence of a steady-state equilibrium. The state variable diverges monotonically to infinity regardless of its initial value, i.e. $\lim_{t\to\infty} y_t = \infty$, $\forall y_0 \in \Re_+$.

---

[5] See Li and Yorke (1975).

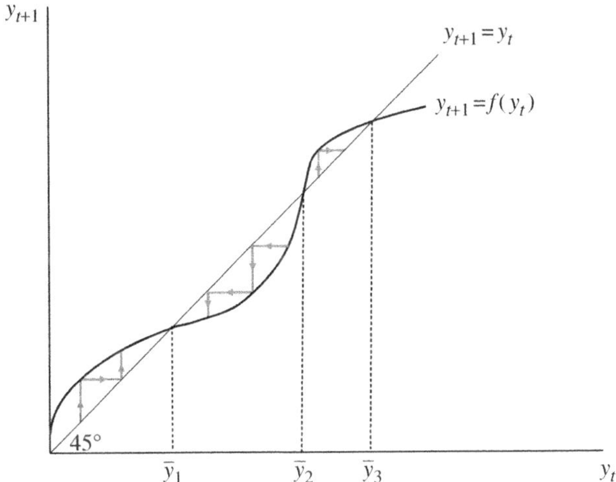

Fig. 1.12. One-Dimensional, First-Order, nonlinear System
Multiple Steady-State Equilibria

1.2.3 Linearization and Local Stability
of Steady-State Equilibria

The evolution of a nonlinear system in the proximity of a steady-state equilibrium can be analyzed based on a linear approximation of this nonlinear motion.

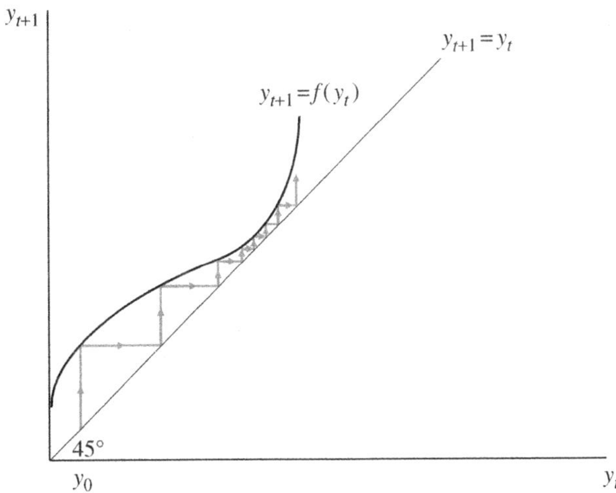

Fig. 1.13. One-Dimensional, First-Order, nonlinear System
Non-Existence of Steady-State Equilibria
$\lim_{t \to \infty} y_t = \infty, \ \forall y_0 \in \Re_+.$

Consider the Taylor expansion of $y_{t+1} = f(y_t)$ around the steady-state level, $\bar{y}$:

$$y_{t+1} = f(y_t) = f(\bar{y}) + f'(\bar{y})(y_t - \bar{y}) + \frac{f''(\bar{y})(y_t - \bar{y})^2}{2!} + \cdots + R_n, \quad (1.16)$$

where R_n is the residual term.

The linearized system around the steady-state equilibrium $\bar{y}$ is therefore

$$y_{t+1} = f(\bar{y}) + f'(\bar{y})(y_t - \bar{y})$$

$$= f'(\bar{y})y_t + f(\bar{y}) - f'(\bar{y})\bar{y} \qquad (1.17)$$

$$= ay_t + b,$$

where $a \equiv f'(\bar{y})$ and $b \equiv f(\bar{y}) - f'(\bar{y})\bar{y}$ are given constants.

Applying the stability results established for the linear system, the linearized system is globally stable if $|a| \equiv |f'(\bar{y})| < 1$. However, since the linear system approximates the behavior of the nonlinear system only in a sufficiently small neighborhood of a steady-state equilibrium, the global analysis of the linearized system provides only a local analysis of the nonlinear difference equation. Thus, the following proposition is established:

Proposition 1.9. *(Necessary and Sufficient Conditions for Local Stability of Steady State Equilibrium)*
The steady-state equilibrium $\bar{y}$ of the dynamical system $y_{t+1} = f(y_t)$ is locally stable if and only if

$$\left| \frac{dy_{t+1}}{dy_t} \bigg|_{\bar{y}} \right| < 1.$$

For example, consider Fig. 1.12 where the dynamical system is characterized by three steady-state equilibria. $f'(\bar{y}_1) < 1$ and $f'(\bar{y}_3) < 1$, and consequently $\bar{y}_1$ and $\bar{y}_3$ are locally stable steady-state equilibria, whereas $f'(\bar{y}_2) > 1$ and consequently $\bar{y}_2$ is an unstable steady-state equilibrium.

Example 1.10.

Consider the nonlinear dynamical system

$$y_{t+1} = 0.5(y_t^2 - 3y_t + 6) \equiv f(y_t), \qquad (1.18)$$

where $y_t \in \Re_+$.

As follows from (1.18), and as depicted in Fig. 1.14, this system has two steady-state equilibria. That is,

$$f(\bar{y}) = \bar{y} \iff \{\bar{y} = 2 \quad \text{or} \quad \bar{y} = 3\}. \qquad (1.19)$$

The linearized system around a steady-state equilibrium $\bar{y}$, as follows from (1.17), is given by

$$y_{t+1} = f'(\bar{y})y_t + f(\bar{y}) - f'(\bar{y})\bar{y}. \qquad (1.20)$$

The linearized system around $\bar{y} = 2$ is therefore

$$y_{t+1} = 0.5(y_t + 2), \qquad (1.21)$$

where

$$\left| \frac{dy_{t+1}}{dy_t} \right|_{\bar{y}=2} = 0.5. \qquad (1.22)$$

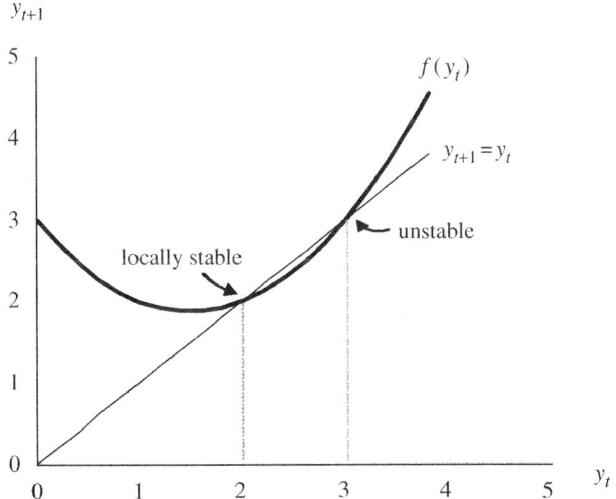

Fig. 1.14. Example of a System with two Steady-State Equilibria

The linearized system around $\bar{y} = 3$ is

$$y_{t+1} = 1.5(y_t - 1), \tag{1.23}$$

where

$$\left| \frac{dy_{t+1}}{dy_t} \right|_{\bar{y}=3} = 1.5. \tag{1.24}$$

Hence, as follows from Proposition 1.9, and as depicted in Fig. 1.14, the dynamical system $y_{t+1} = f(y_t)$ is locally stable around $\bar{y} = 2$ and unstable around $\bar{y} = 3$.

Remark. The evolution of the nonlinear system in the proximity of a steady-state equilibrium, $\bar{y}$, cannot be examined based on the linearized system if $f'(\bar{y}) = 1$. This non-generic case represents a bifurcation point of the dynamical system. Namely, an infinitesimal change in the derivative at the point $\bar{y}$ brings about a *qualitative* change in the nature of the dynamical system (i.e., in the number of steady-state equilibria and their stability).

As depicted in Fig. 1.15, if $f'(\bar{y}) = 1$, then the steady-state equilibrium $\bar{y}$ is neither locally stable nor locally unstable. The state variable converges to the steady-state value, $\bar{y}$, if $y_t < \bar{y}$, whereas it diverges from $\bar{y}$ if $y_t > \bar{y}$. However, the linearized system would falsely suggest that under no initial values the state variable will converge to its steady-state $\bar{y}$.

Example 1.11.

Consider the nonlinear dynamical system

$$y_{t+1} = 0.5(y_t^2 + 1) \equiv f(y_t), \tag{1.25}$$

where $y_t \in \Re_+$.

As follows from (1.25), and as depicted in Fig. 1.15, this system has a single steady-state equilibrium. Namely,

$$f(\bar{y}) = \bar{y} \iff \{\bar{y} = 1\}, \tag{1.26}$$

where $\bar{y} = 1$ is the value of the repeated root of the quadratic equation, $\bar{y}^2 - 2\bar{y} + 1 = 0$.

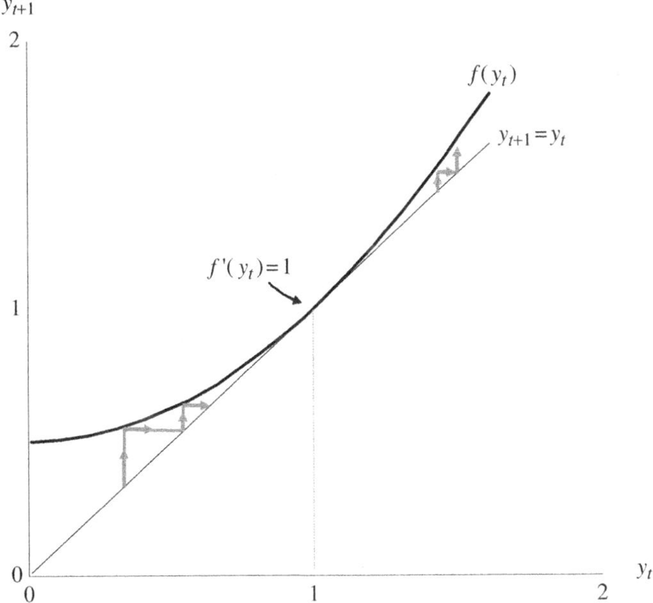

Fig. 1.15. Misleading Linearization
$$f'(\bar{y}) = 1$$

The linearized system around a steady-state equilibrium $\bar{y}$, as follows from (1.17), is given by

$$y_{t+1} = f'(\overline{y})y_t + f(\overline{y}) - f'(\overline{y})\overline{y}, \tag{1.27}$$

and since $\bar{y} = 1$, it is therefore

$$y_{t+1} = y_t, \tag{1.28}$$

where

$$\left| \frac{dy_{t+1}}{dy_t} \bigg|_{\overline{y}=1} \right| = 1. \tag{1.29}$$

Hence, the linearized system is characterized by a continuum of unstable equilibria and in particular, it suggests that the steady-state equilibrium, $\bar{y} = 1$ is unstable, whereas as depicted in Fig. 1.15, the state variable converges to the steady-state value, $\bar{y} = 1$, if $y_t < \bar{y}$, and it diverges from $\bar{y}$ if $y_t > \bar{y}$.

Chapter 4 provides the conditions under which the linearized system can be used in order to examine the properties of a nonlinear dynamical system in the proximity of a steady-state equilibrium.

1.2.4 Global Stability

The *Contraction Mapping Theorem* provides useful sufficient conditions for the existence of a unique steady-state equilibrium and its global stability. These conditions, however, are overly restrictive.

Definition 1.12. *(Contraction Mapping)*
$f(x) : \ \Re \to \Re \ $ *is a contraction mapping if for some* $\ \beta \in (0,1)$

$$\rho(f(x_1), f(x_2)) \leq \beta\rho(x_1, x_2), \qquad \forall x_1, x_2 \in \Re,$$

where $\rho(c, d) \equiv |c - d|$.

Theorem 1.13. *(The Contraction Mapping Theorem)*
If $f : \ \Re \to \Re \ $ *is a contraction mapping then*

- $f(x)$ *has a unique fixed point (steady-state equilibrium), i.e., there exists an* $\bar{x} \in \Re$ *such that*

$$f(\bar{x}) = \bar{x}.$$

- $\forall x_0 \in \Re \ $ *and for* $\ \beta \in (0,1)$,

$$\rho(f^{\{n\}}(x_0), \bar{x}) \leq \beta^n \rho(x_0, \bar{x}) \quad \forall n = 0, 1, 2, 3, \cdots.$$

where $f^{\{n\}}(x_0)$ *is the* n^{th} *iteration of* f *over* x_0.

Proof. See Stokey and Lucas (1989). $\qquad\qquad\square$

Corollary 1.14. *A steady-state equilibrium of the difference equation* $y_{t+1} = f(y_t) \ $ *exists and is unique and globally (asymptotically) stable if*

$f : \ \Re \to \Re \ $ *is a contraction mapping, i.e.*

$$\frac{|f(y_{t+1}) - f(y_t)|}{|y_{t+1} - y_t|} < 1 \qquad \forall t = 0, 1, 2, \cdots, \infty,$$

or if f is differentialble and

$$\left|f'(y_t)\right| < 1 \qquad \forall y_t \in \Re.$$

Thus, as depicted in Fig. 1.16, if over the entire domain, the derivative of $f(y_t)$ is smaller than unity in absolute value, the map $f(y_t)$ has a unique and globally stable steady-state equilibrium. The steady state locus $y_{t+1} = y_t$ intersects with the nonlinear difference equation, $y_{t+1} = f(y_t)$ at the steady-state equilibrium $\bar{y}$. Given y_0, the value of $y_1 = f(y_0)$ can be read from the corresponding value along the curve $y_{t+1} = f(y_t)$. This value of y_1 can be mapped back to the y_t axis via the 45^0 line. Similarly, given y_1, the value of $y_2 = f(y_1)$ can be read from the corresponding value along the curve $y_{t+1} = f(y_t)$ and mapped back to the y_t axis via the 45^0 line. Hence, as depicted in Fig. 1.16, the state variable evolves along the depicted arrows of motion and converges in oscillations to the steady-state equilibrium $\bar{y}$.

For instance consider the one-dimensional, linear, first-order and autonomous system

$$y_{t+1} = f(y_t) = ay_t + b \qquad |a| < 1. \tag{1.30}$$

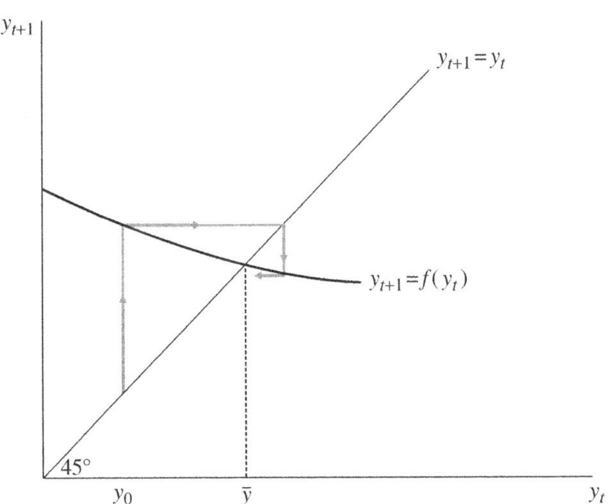

Fig. 1.16. $f(y_t)$ is a Contraction Mapping
Global Stability
$$|f(y_2) - f(y_1)| < |y_2 - y_1| \qquad \forall (y_1, y_2) \in \Re_+$$

As established in Proposition 1.6, since $|a| < 1$, the system has a unique and globally stable steady-state equilibrium. Similarly, based on the *Contraction Mapping Theorem*, as established in Corollary 1.14, there exists a unique and globally stable equilibrium since $|f'(y_t)| = |a| < 1$ $\forall y_t \in \Re$.

2

Multi-Dimensional, First-Order, Linear Systems: Solution

This chapter characterizes the evolution of a vector of state variables in multi-dimensional, first-order linear systems. It develops a method of solution for these multi-dimensional systems, and it characterizes the trajectory of the vector of state variables, in relation to the system's steady-state equilibrium, examining the local and global (asymptotic) stability of this steady-state equilibrium.

Although linear dynamical systems do not necessarily govern the evolution of most of the dynamic phenomena in the universe, they serve as an important benchmark in the analysis of the qualitative properties of nonlinear systems, providing the characterization of the linear approximation of nonlinear systems in proximity of steady-state equilibria.

The characterization of the time path of a multi-dimensional system of interdependent state variables is based on the construction of a time-independent transformation that converts the system into a new dynamical system of either: (a) independent state variables whose evolution can be derived based on the analysis of the one-dimensional case, or (b) partially dependent state variables whose evolution are determined based upon the well established properties of the Jordan Matrix.

The characterization of the trajectories of state variables in multi-dimensional, first-order, linear autonomous systems provides the conceptual foundations for the generalization of the analysis for higher-order, nonlinear, non-autonomous, dynamical systems.

Consider a system of autonomous, first-order, linear difference equations

$$
\begin{bmatrix} x_{1t+1} \\ x_{2t+1} \\ x_{3t+1} \\ \vdots \\ x_{nt+1} \end{bmatrix} = \begin{bmatrix} a_{11} & a_{12} & a_{13} & \cdots & a_{1n} \\ a_{21} & a_{22} & a_{23} & \cdots & a_{2n} \\ a_{31} & a_{32} & a_{33} & \cdots & a_{3n} \\ \vdots & \vdots & \vdots & & \vdots \\ a_{n1} & a_{n2} & a_{n3} & \cdots & a_{nn} \end{bmatrix} \begin{bmatrix} x_{1t} \\ x_{2t} \\ x_{3t} \\ \vdots \\ x_{nt} \end{bmatrix} + \begin{bmatrix} b_1 \\ b_2 \\ b_3 \\ \vdots \\ b_n \end{bmatrix}, \qquad (2.1)
$$

where the initial values of the vector of state variables, $x_0 = (x_{10}, x_{20}, x_{30}, ..., x_{n0})$, are given.

The evolution of the vector of state variables x_t is governed, therefore, by the linear system

$$
x_{t+1} = Ax_t + B, \qquad t = 0, 1, 2, 3, \cdots \qquad (2.2)
$$

where the vector of state variables x_t is an n - dimensional real vector; $x_t \in \Re^n$, A is an $n \times n$ matrix of constant (time-independent) coefficients with elements $a_{ij} \in \Re$, $i, j = 1, 2, ..., n$, and B is an n - dimensional time-independent vector with elements $b_i \in \Re$, $i = 1, 2, ..., n$.

The system is defined as an n - dimensional, first-order, autonomous, linear system of difference equations since it describes the evolution of an n - *dimensional* vector of state variables, x_t, whose value depends in a *linear* and *time-independent* (autonomous) fashion only on the value of the vector in the previous period (first-order).

2.1 Characterization of the Solution

A solution to a multi-dimensional linear system, $x_{t+1} = Ax_t + B$, is a trajectory $\{x_t\}_{t=0}^{\infty}$ of the vector of state variables x_t that satisfies this linear relationship at any point in time. It relates the vector of the state variables at time t, x_t to the vector of initial conditions, x_0, and the set of coefficients embodied in the matrix A and the column vector B. Similarly to the one-dimensional case, the method of iterations generates a pattern that constitutes a general solution rule.

Given the value of the vector of state variables at time 0, x_0, the dynamical system $x_{t+1} = Ax_t + B$ implies that the value of the vector of state variables in subsequent time periods, $1, 2, 3, ...$ is

$$x_1 = Ax_0 + B$$

$$x_2 = Ax_1 + B = A^2x_0 + AB + B$$

$$x_3 = Ax_2 + B = A^3x_0 + A^2B + AB + B \qquad (2.3)$$

$$\vdots \qquad \vdots$$

$$x_t = A^tx_0 + A^{t-1}B + A^{t-2}B + \ldots + AB + B.$$

The value of the vector of state variables in period t is therefore

$$x_t = A^tx_0 + \sum_{i=0}^{t-1} A^iB. \qquad (2.4)$$

It depends on the sum of a geometric series of matrices (rather than of scalars in the one-dimensional case).

Lemma 2.1. *The sum of a geometric series of matrices, $\sum_{i=0}^{t-1} A^i$, whose factor is the matrix A, is*

$$\sum_{i=0}^{t-1} A^i = [I - A^t][I - A]^{-1} \qquad if \quad |I - A| \neq 0.$$

Proof.

$$\sum_{i=0}^{t-1} A^i[I-A] = I + A + A^2 + \ldots + A^{t-1} - [A + A^2 + A^3 + \ldots + A^t] = I - A^t.$$

$$(2.5)$$

Hence, post-multiplication of both sides of the equation by the matrix $[I - A]^{-1}$ establishes the lemma, noting that $[I - A]^{-1}$ exists if and only if $|I - A| \neq 0$. $\qquad\qquad\square$

Using the result in Lemma 2.1, it follows that the solution to the n- dimensional system of linear difference equations is

$$x_t = A^t[x_0 - [I - A]^{-1}B] + [I - A]^{-1}B \qquad if \qquad |I - A| \neq 0, \quad (2.6)$$

where $|I - A| \equiv \det[I - A]$.

The value of the vector of state variables at time t, x_t depends, therefore, on the vector of initial conditions, x_0, and the time-invariant coefficients embodied in the matrix A and the column vector B.

In analogy to the one-dimensional case, the qualitative aspects of the dynamical system are determined by the parameters of the matrix A. These parameters determine whether the dynamical system evolves monotonically or in oscillations, and whether the vector of state variables converges in the long run to a stationary state, diverges asymptotically to plus or minus infinity, or evolves in a periodic orbit.

As will become apparent, in a linear multi-dimensional system the evolution of each state variable may differ qualitatively. Some of the elements of the n - dimensional vector of state variables may converge (monotonically or in oscillations) to a steady-state equilibrium, others may diverge to plus or minus infinity, or may display a periodic orbit. The qualitative examination of a dynamical system requires, therefore, the analysis of the asymptotic behavior of the system as time approaches infinity.

2.2 Existence and Uniqueness of Steady-State Equilibria

Steady-state equilibria provide an essential reference point for a qualitative analysis of the behavior of dynamical systems as time approaches infinity. The qualitative properties of the dynamical system could be assessed based on the examination of the evolution of the vector of state variables in relation to the steady-state equilibria (i.e. the fixed points) of the system.

A *steady-state equilibrium* of this n-dimensional system is a value of the n-dimensional vector of the state variables, x_t that is invariant under further iterations of the dynamical system. Thus, once each of the elements of the vector of state variables is at its steady-state level, the system will not evolve in the absence of exogenous perturbations in the value of the state variable or in the parameters of the matrix A and the vector B.

Definition 2.2. *(A Steady-State Equilibrium)*
A steady-state equilibrium of a linear system of difference equations
$x_{t+1} = Ax_t + B$ *is a vector* $\overline{x} \in \Re^n$ *such that*

$$\overline{x} = A\overline{x} + B.$$

Following the definition, and in analogy to the analysis of the one-dimensional system, there exists a steady-state equilibrium:

$$\overline{x} = [I - A]^{-1}B \qquad \text{if} \qquad |I - A| \neq 0. \tag{2.7}$$

Moreover, in analogy to Proposition 1.3, a steady-state equilibrium is unique if $[I - A]$ is non-singular.

Proposition 2.3. *(Uniqueness of Steady-State Equilibrium)*
A steady-state equilibrium of the system $x_{t+1} = Ax_t + B$ *is unique if and only if*

$$|I - A| \neq 0.$$

The necessary and sufficient condition for the uniqueness of the steady-state equilibrium of the dynamical system is the non-singularity of the matrix $[I - A]$; a condition that is analogous to the requirement that $a \neq 1$ in the one-dimensional case.

The Time Path of x_t

Given the steady-state level of the state variable, x_t, as derived in (2.7), the solution to the difference equation, $x_{t+1} = Ax_t + B$, can be expressed in terms of the deviations of the initial value of the vector of state variables, x_0, from its steady-state value, $\overline{x}$. Substituting (2.7) into (2.6), the solution to the system is

$$x_t = A^t(x_0 - \overline{x}) + \overline{x} \qquad \text{if} \quad |I - A| \neq 0. \tag{2.8}$$

If the matrix A is a diagonal matrix, there exists no interdependence between the different state variables. The matrix A^t is also a diagonal matrix and the evolution of each of the state variables can be analyzed separately according to the method developed for the one-dimensional case in Sect. 1.1.

However, if the matrix A is not a diagonal matrix and there exists interdependence in the evolution of the state variables, a more elaborate method of solution is required. This method generates a time-independent transformation of the system of interdependent state variables into a new dynamical system of either (a) independent state variables that can be analyzed according to the method developed for the one-dimensional case, or (b) interdependent state variables that are governed by a matrix in the *Jordan normal form* and can be examined based on the known limiting properties of this matrix as time approaches infinity.

In particular, as will be elaborated in the next sections, this method transforms a system of interdependent state variables, x_t, into a new system of state variables, y_t, via a time-independent transformation, Q. Namely, this method constructs (based on the eigenvectors of the matrix A) an n x n non-singular matrix, Q, such that

$$x_t = Qy_t + \bar{x}, \tag{2.9}$$

and the evolution of the vector of state variables, y_t, is

$$y_{t+1} = Dy_t, \tag{2.10}$$

where D is an n x n diagonal (or block diagonal) matrix in the *Jordan normal form*. Hence

$$y_t = D^t y_0, \tag{2.11}$$

where, as follows from (2.9), $y_0 = Q^{-1}(x_0 - \bar{x})$.

Moreover, the value of the vector of the state variables, x_t, is a linear, time-independent, transformation of the evolution of y_t, i.e. $x_t = Qy_t + \bar{x}$, whose qualitative behavior will be determined by the qualitative behavior of the vector of state variables y_t. In particular,

$$x_t = QD^t y_0 + \bar{x} = QD^t Q^{-1}(x_0 - \bar{x}) + \bar{x}, \tag{2.12}$$

where Q is a time-invariant matrix and D^t is the Jordan matrix raised to the power t, whose properties are well established. Hence, as time approaches infinity, the vector of state variables, x_t, approaches its steady-state level, $\bar{x}$, if the matrix $QD^t Q^{-1}$ vanishes. Since Q and Q^{-1} are time-independent, it follows that as time approaches infinity, the vector of state variables, x_t, approaches its steady-state level, $\bar{x}$, if D^t vanishes.

If the matrix A has non-repeated (real or complex) eigenvalues, the matrix D is a diagonal matrix and therefore the matrix D^t is a diagonal matrix as well. The time path of each of the elements of the vector y_t is independent of the other and can be examined based on the analysis developed for the one-dimensional case in Sect. 1.1. In particular, D^t vanishes and the vector of state variables, x_t, approaches its steady-state level, $\bar{x}$, if all n elements of the diagonal matrix D are smaller than 1 in absolute value.

If the matrix A has repeated (real or complex) eigenvalues, the matrix D is a block diagonal matrix, and the properties of $\lim_{t \to \infty} D^t$ are well established, permitting a complete characterization of the solution for x_t.

2.3 Examples of Two-Dimensional Systems

This section presents two simple examples of a two-dimensional, first-order, linear, dynamical system that demonstrate the solution method and the qualitative analysis that will be adopted in the case of multidimensional, first-order, linear, dynamical systems. The first example focuses on a discrete dynamical system in which the two state variables evolve independently of one another, demonstrating the direct use of the analysis of the one-dimensional case for the characterization of this system. The second example focuses on a system of interdependent state variables, demonstrating the construction of a time-invariant transformation of this dynamical system into a new dynamical system of independent state variables whose evolution, and therefore the evolution of the original state variables, can be examined on the basis of the insight of the one-dimensional case.

2.3.1 Explicit Solution and Stability Analysis

Consider a system of two-dimensional, first-order, *homogeneous* difference equations[1]

$$x_{t+1} = Ax_t, \tag{2.13}$$

where the value of the vector of state variables in period 0, x_0, is given. As follows from (2.8), or as can be derived directly by the method of iterations, the evolution of the vector of state variables is governed by

$$x_t = A^t x_0. \tag{2.14}$$

If the matrix A is a diagonal matrix as would be the case in Example 2.4, there exists no interdependence between the different state variables. The matrix A^t is also a diagonal matrix, and the evolution of each of the state variables can be analyzed separately according to the method developed for the unidimensional case in Sect. 1.1. A more general form of the matrix A, which implies interdependence across the state variables, would require, however, the construction of a time-independent transformation of the system of interdependent state variables, x_t, into a new dynamical system of independent state variables.

---

[1] The homogeneity of the system is reflected by the fact that $B = 0$ in the general linear system $x_{t+1} = Ax_t + B$, and the system's steady-state equilibrium is therefore the vector 0. As will be established in Sect. 2.5, a non-homogeneous system can be transformed into a homogenous system by shifting the origin of the non-homogeneous system to its steady-state equilibrium. Hence, the use of examples of homogenous systems should be viewed only as a simplifying device.

Example 2.4. (An Uncoupled System)

Consider a system of two-dimensional, first-order, *homogeneous* difference equations, $x_{t+1} = Ax_t$, such that

$$\begin{bmatrix} x_{1t+1} \\ x_{2t+1} \end{bmatrix} = \begin{bmatrix} 2 & 0 \\ 0 & 0.5 \end{bmatrix} \begin{bmatrix} x_{1t} \\ x_{2t} \end{bmatrix}, \tag{2.15}$$

where the initial conditions of the system, $x_0 \equiv [x_{10}, x_{20}]$, are given.

Since the matrix A is a diagonal matrix, the system is uncoupled and the evolution of each of the state variables is independent of the other, i.e. x_{1t+1} depends only on x_{1t}, and x_{2t+1} depends only on x_{2t}.

The solution to the system, $x_t = A^t x_0$, as derived in (2.8), is therefore

$$\begin{bmatrix} x_{1t} \\ x_{2t} \end{bmatrix} = \begin{bmatrix} 2^t & 0 \\ 0 & (0.5)^t \end{bmatrix} \begin{bmatrix} x_{10} \\ x_{20} \end{bmatrix}. \tag{2.16}$$

Since the matrix A^t is a diagonal matrix, the evolution of each state variable is independent of the evolution of the other state variable, and it can be examined in isolation analogously to the solution method developed for the unidimensional case in Sect. 1.1.

In particular,

$$\begin{aligned} x_{1t} &= 2^t x_{10} \\ x_{2t} &= (0.5)^t x_{20}. \end{aligned} \tag{2.17}$$

The steady-state equilibrium of the entire system is therefore

$$(\bar{x}_1, \bar{x}_2) = (0, 0). \tag{2.18}$$

The evolution of each of the state variables as time approaches infinity differs qualitatively. The second state variable, x_{2t}, converges to its steady-state level $\bar{x}_2 = 0$ regardless of its initial value x_{20}. Namely,

$$\lim_{t \to \infty} x_{2t} = \bar{x}_2 = 0, \quad \forall x_{20} \in \Re. \tag{2.19}$$

If $x_{20} > 0$, the value of x_{2t} approaches zero monotonically from the positive quadrants of x_{2t}, whereas if $x_{20} < 0$, it approaches zero monotonically from the negative quadrants of x_{2t}.

In contrast, the first state variable, x_{1t}, diverges to plus or minus infinity, unless the initial position of this state variable is at its steady-state level, $\bar{x}_1 = 0$. Namely, it diverges monotonically to $-\infty$, if $x_{10} < 0$, or to $+\infty$ if $x_{10} > 0$. Namely,

$$\lim_{t \to \infty} x_{1t} = \begin{cases} -\infty & if \quad x_{10} < 0 \\ \bar{x}_1 = 0 & if \quad x_{10} = 0 \\ \infty & if \quad x_{10} > 0. \end{cases} \tag{2.20}$$

As depicted in Fig. 2.1, the steady-state equilibrium, $(\bar{x}_1, \bar{x}_2) = (0, 0)$, is a saddle point.[2] Unless $x_{10} = 0$, the steady-state equilibrium will not be reached, and the system will diverge in its x_{1t} dimension to $-\infty$ or $+\infty$. Namely,

$$\lim_{t \to \infty} x_t = \begin{cases} (-\infty, 0) & if \quad x_{10} < 0 \\ \bar{x} = (0, 0) & if \quad x_{10} = 0 \\ (\infty, 0) & if \quad x_{10} > 0. \end{cases} \tag{2.21}$$

Example 2.5. (A Coupled System)

Consider a coupled system of two-dimensional, first-order, homogeneous difference equations

$$\begin{bmatrix} x_{1t+1} \\ x_{2t+1} \end{bmatrix} = \begin{bmatrix} 1 & 0.5 \\ 1 & 1.5 \end{bmatrix} \begin{bmatrix} x_{1t} \\ x_{2t} \end{bmatrix}, \tag{2.22}$$

where the initial levels of the vector of state variables, $x_0 \equiv [x_{10}, x_{20}]$, is given.

The steady-state equilibrium of this homogeneous system is therefore

---

[2] For the ease of visualization, a single trajectory was drawn in each quadrant. In fact, for each level of x_{1t} there is a continuum of trajectories corresponding to each level of x_{2t}. Moreover, each trajectory is drawn in a continuous manner. The actual trajectories, however, are sequences of discrete points that lie on this continuous trajectory.

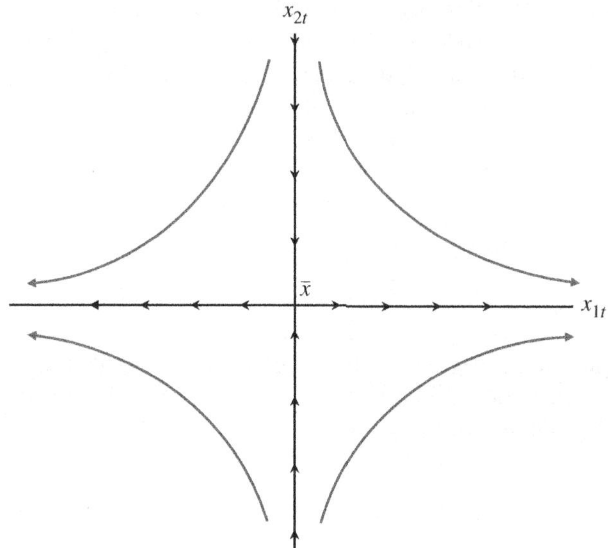

Fig. 2.1. An Uncoupled System
A Saddle

$$(\bar{x}_1, \bar{x}_2) = (0, 0). \tag{2.23}$$

The system cannot be directly uncoupled since the two variables, x_{1t} and x_{2t}, are interdependent. Thus a different solution method is required. The solution technique converts the coupled system (via a time-invariant matrix) into a new system of coordinates in which the dynamical system is uncoupled and is therefore solvable with the method of analysis described for the one-dimensional case in Sect. 1.1.

As is established formally in the next sections, the new system of coordinates is formed by the lines spanned by each of the eigenvectors of the matrix A. Moreover, the time-invariant transformation of the vector of state variables, x_t, into the new system of coordinates, is based on a matrix whose columns are the eigenvectors of the matrix A. Hence the solution method involves the derivation of the eigenvalues and the eigenvectors of the matrix A.

The Eigenvalues of the Matrix A

The eigenvalues of the matrix A are obtained as a solution to the equation

$$|A - \lambda I| = 0, \tag{2.24}$$

where $|A - \lambda I|$ is the determinant of the matrix $[A - \lambda I]$ and I is the identity matrix. In the two-dimensional case, the eigenvalues, λ_1 and λ_2, are therefore obtained as a solution to the equation

$$\begin{vmatrix} a_{11} - \lambda & a_{21} \\ a_{12} & a_{22} - \lambda \end{vmatrix} = 0. \qquad (2.25)$$

The implied characteristic polynomial is

$$c(\lambda) \equiv \lambda^2 - tr A\lambda + \det A = 0, \qquad (2.26)$$

and the eigenvalues are determined by the solution to

$$\begin{cases} \lambda_1 + \lambda_2 = & tr A \\ \lambda_1 \lambda_2 & = \det A. \end{cases} \qquad (2.27)$$

In light of (2.22), it follows that

$$\begin{cases} \lambda_1 + \lambda_2 = 2.5 \\ \lambda_1 \lambda_2 = 1, \end{cases} \qquad (2.28)$$

and therefore $\lambda_1 = 2$ and $\lambda_2 = 0.5$.

The Eigenvectors of the Matrix A

f_1 and f_2, the eigenvectors of the matrix A, that are associated with the eigenvalues λ_1 and λ_2, are obtained as a solution to the equations

$$\begin{aligned} [A - \lambda I]f_1 = 0 & \quad for \quad f_1 \neq 0 \\ [A - \lambda I]f_2 = 0 & \quad for \quad f_2 \neq 0, \end{aligned} \qquad (2.29)$$

where $f_i = (f_{i1}, f_{i2})'$ for $i = 1, 2$. Hence, it follows from (2.22) that the eigenvector associated with the eigenvalue $\lambda_1 = 2$ is determined by the solution to the system of equations

$$\begin{bmatrix} -1 & 0.5 \\ 1 & -0.5 \end{bmatrix} \begin{bmatrix} f_{11} \\ f_{12} \end{bmatrix} = \begin{bmatrix} 0 \\ 0 \end{bmatrix}, \qquad (2.30)$$

whereas that associated with $\lambda_2 = 0.5$ is determined by the solution
to the system of equations

$$\begin{bmatrix} 0.5 & 0.5 \\ 1 & 1 \end{bmatrix} \begin{bmatrix} f_{22} \\ f_{21} \end{bmatrix} = \begin{bmatrix} 0 \\ 0 \end{bmatrix}. \tag{2.31}$$

Thus, the first eigenvector is determined by the equation

$$f_{12} = 2f_{11}, \tag{2.32}$$

whereas the second eigenvector is given by the equation

$$f_{22} = -f_{21}. \tag{2.33}$$

The eigenvectors f_1 and f_2 are therefore

$$f_1 = \begin{bmatrix} 1 \\ 2 \end{bmatrix}$$

$$\tag{2.34}$$

$$f_2 = \begin{bmatrix} 1 \\ -1 \end{bmatrix},$$

or any scalar multiplication of these vectors.

**The Construction of a New System of Coordinates
that Spans $\Re^2$**

Since f_1 and f_2 are linearly independent, they span $\Re^2$. Namely,
for all $x_t \in \Re^2$ there exists $y_t \equiv (y_{1t}, y_{2t}) \in \Re^2$ such that

$$x_t = f_1 y_{1t} + f_2 y_{2t}. \tag{2.35}$$

Hence, as follows from the values of the eigenvectors given in (2.34)

$$\begin{bmatrix} x_{1t} \\ x_{2t} \end{bmatrix} = \begin{bmatrix} 1 & 1 \\ 2 & -1 \end{bmatrix} \begin{bmatrix} y_{1t} \\ y_{2t} \end{bmatrix}. \tag{2.36}$$

Namely, every $x_t = (x_{1t}, x_{2t})' \in \Re^2$ can be expressed in terms of the
new system of coordinates, $(y_{1t}, y_{2t}) \in \Re^2$.

In particular, there exists a time-independent matrix Q such that

$$x_t = Qy_t. \tag{2.37}$$

Since f_1 and f_2 are linearly independent, Q is a non-singular matrix, Q^{-1} therefore exists, and y_t can be expressed in terms of the original system of coordinates (x_{1t}, x_{2t}). That is,

$$y_t = Q^{-1}x_t. \tag{2.38}$$

In particular,

$$\begin{bmatrix} y_{1t} \\ y_{2t} \end{bmatrix} = -\frac{1}{3}\begin{bmatrix} -1 & -1 \\ -2 & 1 \end{bmatrix}\begin{bmatrix} x_{1t} \\ x_{2t} \end{bmatrix}, \tag{2.39}$$

and therefore,

$$y_{1t} = \tfrac{1}{3}(x_{1t} + x_{2t})$$
$$y_{2t} = \tfrac{1}{3}(2x_{1t} - x_{2t}). \tag{2.40}$$

Thus,

$$y_{1t} = 0 \Leftrightarrow x_{2t} = -x_{1t}$$
$$y_{2t} = 0 \Leftrightarrow x_{2t} = 2x_{1t}. \tag{2.41}$$

The geometric place of the new system of coordinates is given therefore by (2.41).

As depicted in Fig. 2.2, the geometric place of all pairs (x_{1t}, x_{2t}) such that $x_{2t} = -x_{1t}$, is the y_{2t} axis (along which $y_{1t} = 0$) and the geometric place of all pairs (x_{1t}, x_{2t}) such that $x_{2t} = 2x_{1t}$ is the y_{1t} axis (along which $y_{2t} = 0$).

The axes of the new system of coordinates (y_{1t}, y_{2t}) are therefore the lines spanned by the eigenvectors f_1 and f_2, respectively, as depicted in Fig. 2.2.

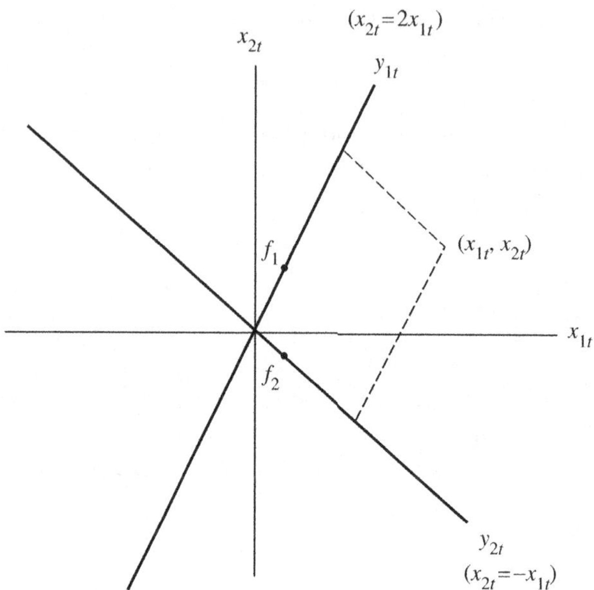

Fig. 2.2. The New System of Coordinates
The Representation of $(x_{1t},\ x_{2t})$ in the $(y_{1t},\ y_{2t})$ space

The Independence of the Evolution of the State Variables y_{1t} and y_{2t}

As follows from (2.38), the value of the vector of state variables y_{t+1} can be expressed as a time-invariant function of the value of the vector of state variables x_{t+1}. In particular,

$$y_{t+1} = Q^{-1}x_{t+1}. \tag{2.42}$$

Hence, since the evolution of the vector of state variables x_{t+1} is given by $x_{t+1} = Ax_t$, it follows that

$$y_{t+1} = Q^{-1}Ax_t. \tag{2.43}$$

Moreover, as established in (2.37), the value of the vector of state variables, x_t, can be expressed in terms of the new system of coordinates, (y_{1t}, y_{2t}). In particular, $x_t = Qy_t$, and therefore

$$y_{t+1} = Q^{-1}AQy_t. \tag{2.44}$$

Thus,

$$y_{t+1} \equiv Dy_t, \tag{2.45}$$

where $D \equiv Q^{-1}AQ$.

As follows from (2.36) and (2.39),

$$D = Q^{-1}AQ = -\frac{1}{3}\begin{bmatrix} -1 & -1 \\ -2 & 1 \end{bmatrix}\begin{bmatrix} 1 & 0.5 \\ 1 & 1.5 \end{bmatrix}\begin{bmatrix} 1 & 1 \\ 2 & -1 \end{bmatrix}$$
$$= \begin{bmatrix} 2 & 0 \\ 0 & 0.5 \end{bmatrix}. \tag{2.46}$$

Namely, the matrix D is a diagonal matrix.

As is evident, the eigenvalues of the matrix A, $\lambda_1 = 2$ and $\lambda_2 = 0.5$ are the diagonal elements of the matrix D, i.e.

$$D = \begin{bmatrix} \lambda_1 & 0 \\ 0 & \lambda_2 \end{bmatrix}. \tag{2.47}$$

Thus, the evolution of each of the elements of the vector of state variables, y_t, is independent of the evolution of the other state variable, and its time path can be determined by the method of solution developed for the unidimensional case in Sect. 1.1.

The Solution for y_t

The evolution of the vector of new state variables, y_t, is given by

$$y_{t+1} = \begin{bmatrix} 2 & 0 \\ 0 & 0.5 \end{bmatrix} y_t. \tag{2.48}$$

Since the system is uncoupled, it follows from the methods of iterations that

$$y_t = D^t y_0, \tag{2.49}$$

namely,

$$y_{1t} = 2^t \, y_{10}$$

$$y_{2t} = (0.5)^t \, y_{20}. \tag{2.50}$$

The value of the vector of new state variables in period 0, y_0, is not given. Nevertheless, it is determined uniquely by the values of the original vector state variables in period 0, x_0. In particular, as follows from (2.38), $y_0 = Q^{-1}x_0$.

The steady-state equilibrium of the system $y_{t+1} = Dy_t$ is a vector $\bar{y} \in \Re^2$ such that $\bar{y} = D\bar{y}$. The steady-state equilibrium of the new system is therefore

$$\bar{y} \equiv (\bar{y}_1, \bar{y}_2)' = (0,0)'. \tag{2.51}$$

The steady-state equilibrium $\bar{y} = (0,0)$ is *unique* since $[I - D]$ is non-singular, i.e.

$$|I - D| = \begin{vmatrix} -1 & 0 \\ 0 & 0.5 \end{vmatrix} = -0.5 \neq 0. \tag{2.52}$$

The second state variable, y_{2t}, converges to its steady-state level $\bar{y}_2 = 0$, regardless of its initial value of y_{20}. Namely,

$$\lim_{t \to \infty} y_{2t} = \bar{y}_2 = 0, \quad \forall y_{20} \in \Re. \tag{2.53}$$

The first state variable, y_{1t}, diverges to plus or minus infinity, unless the initial position of this state variable is at its steady-state level, $\bar{y}_1 = 0$. That is,

$$\lim_{t \to \infty} y_{1t} = \begin{cases} -\infty & if \quad y_{10} < 0 \\ \bar{y}_1 = 0 & if \quad y_{10} = 0 \\ +\infty & if \quad y_{10} > 0. \end{cases} \tag{2.54}$$

As depicted in Fig. 2.3, the steady-state equilibrium $(\bar{y}_1, \bar{y}_2) = (0,0)$ is a saddle point. Namely, unless $y_{10} = 0$, the steady-state equilibrium

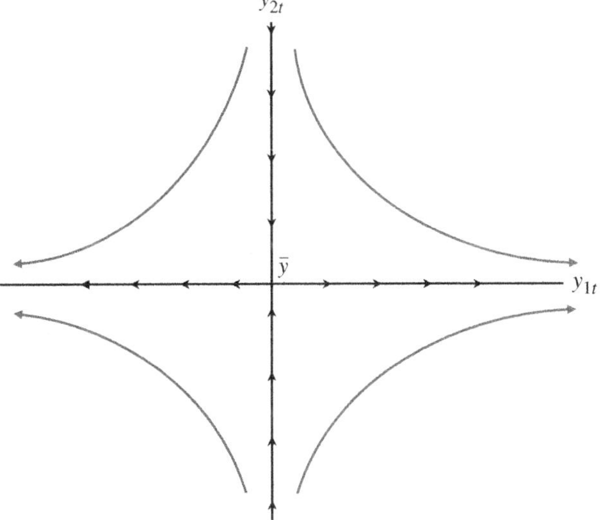

Fig. 2.3. The Evolution of y_t

will not be reached, and the system will diverge in one of its dimensions
to either plus or minus infinity. Thus,

$$\lim_{t \to \infty} y_t = \begin{cases} (-\infty, 0) & if \quad y_{10} < 0 \\ (0,0) & if \quad y_{10} = 0 \\ (\infty, 0) & if \quad y_{10} > 0. \end{cases} \tag{2.55}$$

The Solution for x_t

The position of the vector of state variables, x_t, can be expressed
in term of the new system of coordinates, (y_{1t}, y_{2t}). In particular, as
established in (2.37), $x_t = Qy_t$, i.e.

$$\begin{bmatrix} x_{1t} \\ x_{2t} \end{bmatrix} = \begin{bmatrix} 1 & 1 \\ 2 & -1 \end{bmatrix} \begin{bmatrix} y_{1t} \\ y_{2t} \end{bmatrix}. \tag{2.56}$$

Hence, a time-invariant transformation of the explicit solution for the time path of the vector of new state variables, y_t, provides an explicit solution for the time path of the original vector of state variables, x_t.

As follows from (2.50),

$$
\begin{bmatrix} x_{1t} \\ x_{2t} \end{bmatrix} = \begin{bmatrix} 1 & 1 \\ 2 & -1 \end{bmatrix} \begin{bmatrix} 2^t y_{10} \\ (0.5)^t y_{20} \end{bmatrix} = \begin{bmatrix} 2^t y_{10} + (0.5)^t y_{20} \\ 2^{t+1} y_{10} - (0.5)^t y_{20} \end{bmatrix}, \quad (2.57)
$$

where the initial value $y_0 = Q^{-1} x_0$. Hence, it follows from (2.40) that

$$
\begin{bmatrix} y_{10} \\ y_{20} \end{bmatrix} = -\frac{1}{3} \begin{bmatrix} -1 & -1 \\ -2 & 1 \end{bmatrix} \begin{bmatrix} x_{10} \\ x_{20} \end{bmatrix}, \quad (2.58)
$$

and therefore,

$$
y_{10} = \tfrac{1}{3}(x_{10} + x_{20})
$$
$$
y_{20} = \tfrac{1}{3}(2x_{10} - x_{20}). \quad (2.59)
$$

The time path of x_t and its qualitative properties are therefore uniquely determined by the system's initial conditions, (x_{10}, x_{20}), and the eigenvalues of the matrix A.

$$
\begin{bmatrix} x_{1t} \\ x_{2t} \end{bmatrix} = \begin{bmatrix} \frac{2^t}{3}(x_{10} + x_{20}) + \frac{(0.5)^t}{3}(2x_{10} - x_{20}) \\ \frac{2^{t+1}}{3}(x_{10} + x_{20}) - \frac{(0.5)^t}{3}(2x_{10} - x_{20}) \end{bmatrix}. \quad (2.60)
$$

The phase diagram of the original system is obtained by placing the phase diagram that describes the evolution of y_t, relative to the new system of coordinates (y_1, y_2), in the plane (x_1, x_2), as depicted in Fig. 2.4.

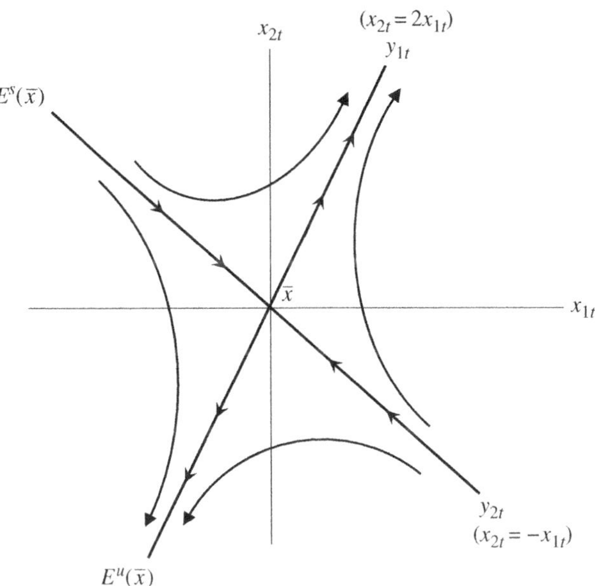

Fig. 2.4. The Evolution of x_t
A Saddle

Stability of the Steady-State Equilibrium $\bar{x}$

A steady-state equilibrium of the system $x_{t+1} = Ax_t$ is a vector $\bar{x} \in \Re^2$ such that $\bar{x} = A\bar{x}$. Hence, it follows from (2.60) that $\bar{x}$ exists and is given by $\bar{x} = (0,0)$. Moreover, $\bar{x} = (0,0)$ is *unique* since the matrix $[I - A]$ is non-singular, i.e.

$$|I - A| = \begin{vmatrix} 0 & -0.5 \\ -1 & -0.5 \end{vmatrix} = -0.5 \neq 0. \tag{2.61}$$

As follows from (2.60), and as depicted in Fig. 2.4,

$$\lim_{t \to \infty} x_t = \bar{x} \quad \Leftrightarrow \quad x_{20} = -x_{10}, \tag{2.62}$$

and the steady-state equilibrium $\bar{x} = 0$ is a saddle point. Namely, the vector of the original state variables, x_t, converges to its steady-state value $\bar{x}$ if and only if the initial values of this vector are placed on the y_{2t} axis, i.e. along the geometric place of all pairs of (x_{1t}, x_{2t}) such that $x_{20} = -x_{10}$.

2.3.2 Stability Analysis Without an Explicit Solution

A. Construction of Phase Diagrams Without an Explicit Solution

The derivation of a phase diagram for a two-dimensional, first-order linear system does not require an explicit characterization of the evolution of the vector of state variables. The phase diagram can be derived via a characterization of the map of forces that operate on the vector of state variables in any position in the two-dimensional plane.

The construction of the phase diagram requires the identification of the geometric place under which a given state variable is in a steady state, as well as the characterization of the forces that operate on this state variable once it deviates from its steady-state value.

Consider Example 2.5 where

$$\begin{bmatrix} x_{1t+1} \\ x_{2t+1} \end{bmatrix} = \begin{bmatrix} 1 & 0.5 \\ 1 & 1.5 \end{bmatrix} \begin{bmatrix} x_{1t} \\ x_{2t} \end{bmatrix}. \tag{2.63}$$

The system can be rewritten in a slightly different manner, in terms of changes in the values of each of the state variables between time t and time $t+1$. Let Δx_{it} be the change in the value of the ith state variable, $i = 1, 2$, from period t to period $t+1$.

$$\Delta x_{1t} \equiv x_{1t+1} - x_{1t} = 0.5x_{2t}$$
$$\Delta x_{2t} \equiv x_{2t+1} - x_{2t} = x_{1t} + 0.5x_{2t}. \tag{2.64}$$

Clearly, at a steady-state equilibrium, neither x_{1t} nor x_{2t} changes over time and therefore $\Delta x_{1t} = \Delta x_{2t} = 0$.

Let '$\Delta x_{1t} = 0$' be the geometric place of all pairs of (x_{1t}, x_{2t}) such that x_{1t} is in a steady state, and let '$\Delta x_{2t} = 0$' be the geometric place of all pairs (x_{1t}, x_{2t}) such that x_{2t} is in a steady state. Namely,

$$'\Delta x_{1t} = 0' \equiv \{(x_{1t}, x_{2t})|\ x_{1t+1} - x_{1t} = 0\}$$
$$'\Delta x_{2t} = 0' \equiv \{(x_{1t}, x_{2t})|\ x_{2t+1} - x_{2t} = 0\}. \tag{2.65}$$

It follows from (2.64) and (2.65) that

$$\Delta x_{1t} = 0 \quad \Leftrightarrow \quad x_{2t} = 0$$

$$\Delta x_{2t} = 0 \quad \Leftrightarrow \quad x_{2t} = -2x_{1t}.$$

$$(2.66)$$

Thus, as depicted in Fig. 2.5, the geometric locus of all pairs of (x_{1t}, x_{2t}) such that '$\Delta x_{1t} = 0$' is the entire x_{1t} axis, whereas the geometric place of all pairs of (x_{1t}, x_{2t}) such that '$\Delta x_{2t} = 0$' is given by the equation $x_{2t} = -2x_{1t}$.

The geometric place under which the two loci, '$\Delta x_{1t} = 0$' and '$\Delta x_{2t} = 0$,' intersect is the steady state equilibrium of the system. As follows from (2.66) and as depicted in Fig. 2.5, the two loci intersect at the point $(0,0)$ and this is the unique steady-state equilibrium of the entire system.

The forces that operate on each of the state variables out of its steady-state equilibrium provide the necessary elements for the derivation of the phase diagram.

As follows from (2.64), as long as $x_{2t} > 0$, the system increases the value of the first state variable, x_{1t}, in the transition from time t to time $t+1$, whereas if $x_{2t} < 0$ the system tends to decrease the value of

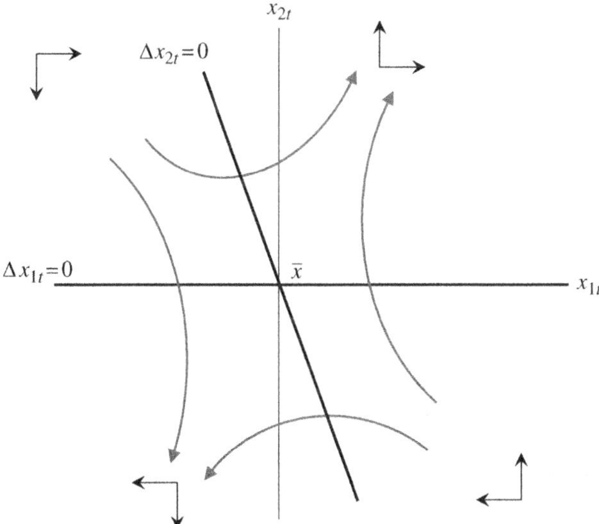

Fig. 2.5. The Construction of a Phase Diagram without Reference to an Explicit Solution

x_{1t} in the transition from time t to time $t + 1$. Hence,

$$\Delta x_{1t} \begin{cases} > 0 & if \quad x_{2t} > 0 \\ \\ < 0 & if \quad x_{2t} < 0. \end{cases} \tag{2.67}$$

Consequently, as depicted in Fig. 2.5, above the x_{1t} axis, the arrows that depict the motion of the first state variable, x_{1t}, are directed rightward, whereas below the x_{1t} axis, the arrows that depict the motion of x_{1t} are directed leftward.

Similarly, if $x_{2t} > -2x_{1t}$, the system tends to increase value of the second state variable, x_{2t}, in the transition from time t to time $t + 1$, whereas if $x_{2t} < -2x_{1t}$, the system tends to decrease the value of x_{2t}. Hence,

$$\Delta x_{2t} \begin{cases} > 0 & if \quad x_{2t} > -2x_{1t} \\ \\ < 0 & if \quad x_{2t} < -2x_{1t}. \end{cases} \tag{2.68}$$

Consequently, as depicted in Fig. 2.5, above the geometric locus $x_{2t} = -2x_{1t}$ (i.e. for pairs of (x_{1t}, x_{2t}) such that $x_{2t} > -2x_{1t}$), the arrows that depict the motion of the second state variable, x_{2t}, are directed upward, whereas below the geometric locus $x_{2t} = -2x_{1t}$ (i.e. for pairs of (x_{1t}, x_{2t}) such that $x_{2t} < -2x_{1t}$), the arrows that depict the motion of x_{2t} are directed downward.

Remark. The motion of state variables in discrete dynamical systems could not be inferred from the arrows of motion relative to the steady-state loci unless the nature of the eigenvalues that characterize the matrix of coefficients of the vector of state variables is verified. In particular, if both eigenvalues are real and positive, each state variable converges or diverges monotonically and the arrows of motion properly describe the motion of the state variables. However, if an eigenvalue is negative, then the dynamical system displays an oscillatory behavior and the motion of the system exhibits oscillations around its corresponding steady-state locus. Similarly, if the eigenvalues are complex, then the dynamical system exhibits a cyclical motion. Hence, in the case of negative or complex eigenvalues, the arrows of motion simply indicate the direction of the oscillations, or the cyclical motion, from each region of the phase diagram.

Since both eigenvalues in Example 2.5 are real and positive, convergence and divergence are monotonic and the arrows of motion approximate the actual motion of the system. The phase diagram is depicted in Fig. 2.5 according to the location of the loci '$\Delta x_{1t} = 0$' and '$\Delta x_{2t} = 0$,' as well as the corresponding arrows of motion.

B. Derivation of a New System of Coordinates Without an Explicit Solution

The derivation of the new system of coordinates is feasible in some cases without an explicit derivation of the solution for the dynamical system.

In particular, if the steady-state equilibrium is a saddle, as is the case in Example 2.5, since the system is linear, convergence to the steady-state equilibrium is along a linear segment that represents the axis of the new system. Moreover, since the system is homogeneous, it follows that the ratio between the two state variables along the convergence path is identical at any point in time. Thus,

$$\frac{x_{2t+1}}{x_{1t+1}} = \frac{x_{2t}}{x_{1t}}, \tag{2.69}$$

along the axis of the new system of coordinates (i.e. along the geometric place from which the vector of state variables converges to the steady-state equilibrium), as well as along the other axis of the new system of coordinates (i.e. along the geometric place from which the vector of state variables converges to the steady-state equilibrium upon backward iterations under the system's law of motion).

Given the structure of the dynamical system in (2.63), it follows that

$$\frac{x_{1t} + 1.5x_{2t}}{x_{1t} + 0.5x_{2t}} = \frac{x_{2t}}{x_{1t}}. \tag{2.70}$$

Hence, the ratio between x_{2t} and x_{1t} along the new system of coordinates (y_{1t}, y_{2t}) is given by the quadratic equation

$$\left(\frac{x_{2t}}{x_{1t}}\right)^2 - \left(\frac{x_{2t}}{x_{1t}}\right) - 2 = 0. \tag{2.71}$$

Based on the solution to this quadratic equation, it follows that the ratios between the two state variables along the new system of coordinates is

$$\frac{x_{2t}}{x_{1t}} = \{2, -1\}. \tag{2.72}$$

These two solutions for this quadratic equation provide the equations of the dashed lines depicted in Fig. 2.5 – lines that are spanned by each eigenvector of the matrix A. They are the two constant ratios that lead into the steady-state equilibrium upon a sufficient number of either forward or backward iterations.

C. Stable and Unstable Eigenspace

Examples 2.4 and 2.5 provide an ideal setting for the introduction of the concepts of a *stable eigenspace* and an *unstable eigenspace*, setting the stage for the introduction of the concepts of stable and unstable manifolds of nonlinear dynamical systems.

The *stable eigenspace*, relative to the steady-state equilibrium $\overline{x}$, is defined as the plane spanned by the eigenvectors associated with eigenvalues of modulus smaller than one. Namely,

$$E^s(\overline{x}) = span \ \{eigenvectors \ associated \ with \ eigenvalues$$

$$of \ modulus < 1\}.$$

In a homogenous two-dimensional autonomous linear system, $x_{t+1} = Ax_t$, the stable eigenspace, $E^s(\overline{x})$, relative to the steady-state equilibrium $\overline{x}$, is

$$E^s(\overline{x}) = \{(x_{1t}, x_{2t})| \lim_{n \to \infty} A^n \begin{pmatrix} x_{1t} \\ x_{2t} \end{pmatrix} = \overline{x}\}. \tag{2.73}$$

Namely, the *stable eigenspace*, is the geometric locus of all pairs (x_{1t}, x_{2t}) that upon a sufficient number of forward iterations are mapped in the limit towards the steady-state equilibrium, $\overline{x}$. The stable eigenspace in Example 2.5 is one-dimensional. It is the line given by the equation $x_{2t} = -x_{1t}$, as depicted in Fig. 2.4.

The *unstable eigenspace*, relative to the steady-state equilibrium $\overline{x}$, is defined as

$$E^u(\overline{x}) = span \ \{eigenvectors \ associated \ with \ eigenvalues$$

$$of \ modulus > 1\}.$$

In a homogeneous two-dimensional linear system, $x_{t+1} = Ax_t$, the unstable eigenspace, $E^u(\overline{x})$, is

$$E^u(\overline{x}) = \{(x_{1t}, x_{2t})| \lim_{n \to \infty} A^{-n} \begin{pmatrix} x_{1t} \\ x_{2t} \end{pmatrix} = \overline{x}\}. \tag{2.74}$$

Namely, the unstable eigenspace is the geometric locus of all pairs (x_{1t}, x_{2t}) that upon a sufficient number of backward iterations are mapped in the limit towards the steady-state equilibrium $\bar{x}$. The unstable eigenspace in Example 2.5 is one-dimensional as well. It is the line given by the equation $x_{2t} = 2x_{1t}$, as depicted in Fig. 2.4.

2.4 Properties of the Jordan Matrix

This section summarizes fundamental propositions from the field of Linear Algebra used in the derivation and qualitative analysis of the evolution of multidimensional discrete dynamical systems.

Consider the multidimensional first-order linear system, $x_{t+1} = Ax_t + B$. If the matrix A is a diagonal matrix, there exists no interdependence between the different state variables. The matrix A^t is also a diagonal matrix, and the evolution of each of the state variables can be analyzed separately according to the method developed for the one-dimensional case in Sect. 1.1.

However, if the matrix A is not a diagonal matrix and there exists interdependence in the evolution of state variables, the characterization of the evolution of this multi-dimensional interdependent dynamical system necessitates a time-independent transformation of the system into either: (a) a new dynamical system of independent state variables that can be analyzed according to the method developed for the one-dimensional case, or (b) a new interdependent system whose interdependence is represented by a matrix in the *Jordan normal form*, and thus can be analyzed on the basis of the well established properties of the nth iteration of the Jordan matrix as n approaches infinity. In particular, this requires the transformation of a system of interdependent state variables, x_t, into a new dynamical system of state variables, y_t, via a time-independent transformation, Q, such that $x_t = Qy_t + \bar{x}$, where $y_{t+1} = Dy_t$, and D is the Jordan matrix, as described below.

Established results in Linear Algebra provide the properties of the matrix of coefficients A that permit the time-invariant transformation into a dynamical system of: (a) independent state variables whose evolution can be characterized based on the unidimensional case examined in Sect. 1.1, or (b) an interdependent system whose evolution is governed by the *Jordan matrix*.

Lemma 2.6 provides the conditions on the matrix of coefficients A under which it is diagonalizable.

Lemma 2.6. *Let* $A = (a_{ij})$ *be an* $n \times n$ *matrix where* $a_{ij} \in \Re$, $i, j = 1, 2, \cdots, n$.

- *If the matrix* A *has* n *distinct real eigenvalues* $\{\lambda_1, \lambda_2, \lambda_3, \cdots, \lambda_n\}$, *then there exists a nonsingular* $n \times n$ *matrix,* Q, *such that*

$$A = QDQ^{-1},$$

where D *is a diagonalized matrix*

$$D = \begin{bmatrix} \lambda_1 & 0 & 0 & \cdots & 0 \\ 0 & \lambda_2 & 0 & \cdots & 0 \\ 0 & 0 & \lambda_3 & \cdots & 0 \\ \vdots & \vdots & \vdots & \ddots & \\ 0 & 0 & 0 & \cdots & \lambda_n \end{bmatrix},$$

and Q *is an invertible* $n \times n$ *matrix whose columns are the eigenvectors of the matrix* A, $\{f_1, f_2, f_3, \ldots f_n\}$, *i.e.*

$$Q = [f_1, f_2, f_3, \ldots f_n].$$

- *If the matrix* A *has* n *repeated real eigenvalues* $\{\lambda, \lambda, \lambda, \lambda, \cdots, \lambda\}$, *then there exists a nonsingular* $n \times n$ *matrix,* Q, *such that*

$$A = QDQ^{-1},$$

where

$$D = \begin{bmatrix} \lambda & 0 & 0 & 0 & \cdots & 0 \\ 1 & \lambda & 0 & 0 & \ddots & 0 \\ 0 & 1 & \lambda & 0 & \ddots & 0 \\ 0 & 0 & 1 & \ddots & \ddots & 0 \\ 0 & 0 & 0 & \ddots & \lambda & 0 \\ 0 & 0 & 0 & \cdots & 1 & \lambda \end{bmatrix}.$$

Proof. See Hirsch and Smale (1974).   □

Lemma 2.7. *If the matrix A has $n/2$ pairs of distinct complex eigenvalues,*

$$\{\mu_1, \overline{\mu}_1, \mu_2, \overline{\mu}_2, \cdots, \mu_{n/2}, \overline{\mu}_{n/2}\}, \quad where$$

$$\mu_j \equiv \alpha_j + \beta_j i,$$
$$\overline{\mu}_j \equiv \alpha_j - \beta_j i,$$

and $i \equiv \sqrt{-1}$, then there exists a nonsingular $n \times n$ matrix, V, such that

$$A = VDV^{-1},$$

where

$$D = \begin{bmatrix} \alpha_1 & -\beta_1 & 0 & 0 & \dots & \dots & 0 & 0 \\ \beta_1 & \alpha_1 & 0 & 0 & \dots & \dots & 0 & 0 \\ 0 & 0 & \alpha_2 & -\beta_2 & \ddots & \ddots & 0 & 0 \\ 0 & 0 & \beta_2 & \alpha_2 & & \ddots & 0 & 0 \\ \vdots & \vdots & \ddots & \ddots & \ddots & \ddots & 0 & 0 \\ \vdots & \vdots & \ddots & \ddots & \ddots & \ddots & 0 & 0 \\ 0 & 0 & 0 & 0 & \dots & \dots & \alpha_{n/2} & -\beta_{n/2} \\ 0 & 0 & 0 & 0 & \dots & \dots & \beta_{n/2} & \alpha_{n/2} \end{bmatrix}.$$

- *If a matrix A has $n/2$ pairs of repeated complex eigenvalues, $\{\mu, \overline{\mu}, \mu, \overline{\mu}, \cdots, \mu, \overline{\mu}\}$, where*

$$\mu \equiv \alpha + \beta i,$$
$$\overline{\mu} \equiv \alpha - \beta i,$$

and $i \equiv \sqrt{-1}$, then there exists a nonsingular $n \times n$ matrix V such that $A = VDV^{-1}$, where

$$D = \begin{bmatrix} \alpha & -\beta & 0 & 0 & 0 & 0 & 0 & 0 \\ \beta & \alpha & 0 & 0 & 0 & 0 & 0 & 0 \\ 1 & 0 & \alpha & -\beta & 0 & 0 & 0 & 0 \\ 0 & 1 & \beta & \alpha & 0 & 0 & 0 & 0 \\ \vdots & \vdots & \ddots & \ddots & \ddots & \ddots & \vdots & \vdots \\ \vdots & \vdots & \ddots & \ddots & \ddots & \ddots & \vdots & \vdots \\ 0 & 0 & 0 & 0 & 1 & 0 & \alpha & -\beta \\ 0 & 0 & 0 & 0 & 0 & 1 & \beta & \alpha \end{bmatrix}.$$

Proof. See Hirsch and Smale (1974). $\qquad\qquad\square$

The following lemma generalizes the results of the pervious one, for the case in which the matrix A may have a mixture of real distinct eigenvalues, repeated real eigenvalues, distinct complex eigenvalues, and repeated complex eigenvalues.

Lemma 2.8. *Let* A *be an* $n \times n$ *matrix where* $a_{ij} \in \Re$, $i,j = 1,2,\cdots,n$. *Then, there exists an* $n \times n$ *nonsingular matrix* V *such that*

$$A = VDV^{-1},$$

and D is the Jordan normal form that corresponds to the matrix A, i.e.

$$D = \begin{bmatrix} D_1 & 0 & 0 & 0 & \cdots & 0 \\ 0 & D_2 & 0 & 0 & \ddots & 0 \\ 0 & 0 & \ddots & 0 & \ddots & 0 \\ 0 & 0 & 0 & D_h & \ddots & 0 \\ \vdots & \ddots & \ddots & \ddots & \ddots & \\ 0 & 0 & 0 & 0 & \cdots & D_m \end{bmatrix}.$$

- *For each distinct real eigenvalue, λ_h, of the matrix A*

$$D_h = \lambda_h.$$

- *For repeated real eigenvalues $\{\lambda, \lambda, ...\lambda\}$ of the matrix A*

$$D_h = \begin{bmatrix} \lambda & 0 & 0 & \cdots & 0 \\ 1 & \lambda & 0 & \ddots & 0 \\ 0 & 1 & \lambda & \ddots & 0 \\ \vdots & \ddots & \ddots & \ddots & 0 \\ 0 & 0 & 0 & \cdots & \lambda \end{bmatrix}.$$

- *For a distinct pair of complex eigenvalues, $\{\mu_h, \overline{\mu}_h\}$, of the matrix A, where $\mu_h \equiv \alpha_h + \beta_h i$ and $\overline{\mu}_h \equiv \alpha_h - \beta_h i$,*

$$D_h = \begin{bmatrix} \alpha_h & -\beta_h \\ \beta_h & \alpha_h \end{bmatrix}.$$

- *For pairs of repeated complex eigenvalues, $\{\mu, \overline{\mu}, \mu, \overline{\mu}, \cdots, \mu, \overline{\mu}\}$, of the matrix A, where $\mu \equiv \alpha + \beta i$, and $\overline{\mu} \equiv \alpha - \beta i$,*

$$D_h = \begin{bmatrix} \alpha & -\beta & 0 & 0 & 0 & 0 & 0 & 0 \\ \beta & \alpha & 0 & 0 & 0 & 0 & 0 & 0 \\ 1 & 0 & \alpha & -\beta & 0 & 0 & 0 & 0 \\ 0 & 1 & \beta & \alpha & 0 & 0 & 0 & 0 \\ \vdots & \vdots & \ddots & \ddots & \ddots & \ddots & \vdots & \vdots \\ \vdots & \vdots & \ddots & \ddots & \ddots & \ddots & \vdots & \vdots \\ 0 & 0 & 0 & 0 & 1 & 0 & \alpha & -\beta \\ 0 & 0 & 0 & 0 & 0 & 1 & \beta & \alpha \end{bmatrix}.$$

Proof. See Hirsch and Smale (1974). $\square$

2.5 Representation of the System in the Jordan Normal Form

The characterization of the evolution of a multidimensional system of interdependent state variables necessitates the construction of a time-independent transformation of the system into a new dynamical system of either (a) independent state variables whose evolution can be derived based on the analysis of the unidimensional case, or (b) partially dependent state variables whose evolution is determined by well established properties of the Jordan matrix.

This section presents the solution to a multidimensional first-order linear discrete system in terms of the Jordan normal form, D. It transforms in a time-independent fashion the original dynamical system into a new one whose evolution is governed by the Jordan matrix. Based upon the well established properties of the Jordan matrix, this representation permits the examination of the properties of the system as time approaches infinity.

Consider a system of autonomous, first-order, linear difference equations

$$x_{t+1} = Ax_t + B, \qquad t = 0, 1, 2, 3, \cdots \tag{2.75}$$

The transformation of a system of interdependent state variables, x_t, into a new system of independent state variables, y_t, whose evolution is governed by the Jordan matrix, D, requires the construction of two matrices: a time-invariant matrix, Q, based upon the eigenvectors of the matrix A, and the Jordan normal form D, using the eigenvalues of the matrix A.

2.5.1 Transformation of Non-Homogeneous Systems into Homogeneous Ones

A non-homogeneous dynamical system can be transformed into a homogenous dynamical system by re-positioning the origin of the axis of the non-homogeneous system at its steady-state equilibrium.

Proposition 2.9. *(Transformation of Non-Homogeneous Systems into Homogeneous Ones)*
A non-homogeneous system of first-order linear difference equations

$$x_{t+1} = Ax_t + B,$$

can be transformed into a homogeneous system of first-order linear difference equations

$$z_{t+1} = Az_t,$$

where $z_t \equiv x_t - \overline{x}$ is a state variable that characterizes the evolution of the deviation of x_t from its steady state value $\overline{x} = [I - A]^{-1}B$.

Proof: Let

$$z_t \equiv x_t - \overline{x}. \tag{2.76}$$

It follows that

$$z_{t+1} = x_{t+1} - \bar{x}. \tag{2.77}$$

Since $x_{t+1} = Ax_t + B$,

$$\begin{aligned} z_{t+1} &= Ax_t + B - \bar{x} \\ &= A(z_t + \overline{x}) + B - \overline{x} \\ &= Az_t - [I - A]\overline{x} + B. \end{aligned} \tag{2.78}$$

Hence, since $\overline{x} = [I - A]^{-1}B$,

$$z_{t+1} = Az_t. \tag{2.79}$$

$\square$

2.5.2 The Solution in Terms of the Jordan Normal Form

Proposition 2.10. *(Solution in Term of the Jordan Normal Form) The evolution of the vector of state variables, x_t, in a system of non-homogeneous first-order linear difference equations*

$$x_{t+1} = Ax_t + B,$$

is given by

$$x_t = QD^tQ^{-1}(x_0 - \overline{x}) + \overline{x},$$

where D is a matrix in the Jordan normal form that corresponds to A, x_0 is the system's initial condition, and $\overline{x} = [I - A]^{-1}B$ is the steady-state equilibrium of the system.

Proof: Let $z_t \equiv x_t - \bar{x}$. It follows from Proposition 2.9 that

$$z_{t+1} = Az_t. \tag{2.80}$$

As follows from Lemma 2.8 there exists a non-singular matrix Q such that

$$A = QDQ^{-1}, \tag{2.81}$$

where D is the Jordan matrix . Thus,

$$z_{t+1} = QDQ^{-1}z_t. \tag{2.82}$$

Let

$$y_t \equiv Q^{-1}z_t. \tag{2.83}$$

Pre-multiplying both sides of (2.82) by Q^{-1}, it follows that

$$y_{t+1} = Dy_t. \tag{2.84}$$

Using the method of iterations, the evolution of the new state variable, y_t is given by

$$y_t = D^t y_0. \tag{2.85}$$

Hence, since $y_0 \equiv Q^{-1}z_0$ and $z_0 = x_0 - \bar{x}$, it follows that

$$y_t = D^t Q^{-1}(x_0 - \bar{x}). \tag{2.86}$$

Furthermore, since $Q^{-1}z_t = y_t$, it follows that $z_t = Qy_t$, and therefore $z_t \equiv x_t - \bar{x} = Qy_t$. Hence,

$$x_t = Qy_t + \bar{x}, \tag{2.87}$$

and therefore

$$x_t = QD^t Q^{-1}(x_0 - \bar{x}) + \bar{x}. \tag{2.88}$$

$\square$

Hence the evolution of the system depends on the properties of the Jordan matrix, D. These properties, in turn, depend on the eigenvalues of the matrix A.

3

Multi-Dimensional, First-Order, Linear Systems: Characterization

This chapter characterizes the trajectory of a vector of state variables in multi-dimensional, first-order, linear dynamical systems. It examines the trajectories of these systems when the matrix of coefficients has real eigenvalues and the vector of state variables converges or diverges in a monotonic or oscillatory fashion towards or away from a steady-state equilibrium that is characterized by either a saddle point or a stable or unstable (improper) node. In addition, it examines the trajectories of these linear dynamical systems when the matrix of coefficients has complex eigenvalues and the system is therefore characterized by a spiral sink, a spiral source, or a periodic orbit.

The qualitative pattern of a multi-dimensional, linear dynamical system, $x_{t+1} = Ax_t + B$, depends upon the nature of the matrix of coefficients A. The trajectory of the system depends on whether the matrix of coefficients A has: (a) distinct real eigenvalues, (b) repeated real eigenvalues, (c) distinct complex eigenvalues, or (d) repeated complex eigenvalues. As established in Sect. 2.4, the Jordan matrix has a different representation under each of these categories, and the evolution of the system is affected accordingly.

3.1 Distinct Real Eigenvalues

3.1.1 Characterization of the Solution

Consider the multi-dimensional, linear dynamical system

$$x_{t+1} = Ax_t + B, \qquad t = 0, 1, 2, 3, \cdots \tag{3.1}$$

where the vector of state variables, x_t, is an n - dimensional real vector; $x_t \in \Re^n$, A is an $n \times n$ matrix of constant (time-independent) coefficients with elements, $a_{ij} \in \Re$, $i, j = 1, 2, ..., n$, and B is an n - dimensional time-independent vector with elements $b_i \in \Re$, $i = 1, 2, ..., n$.

Suppose that the n x n matrix of coefficients, A, has n distinct real eigenvalues $\{\lambda_1, \lambda_2, \cdots, \lambda_n\}$. Suppose further that $|I - A| \neq 0$. As established in Lemma 2.6 and (2.88), there exists a time-invariant transformation of the vector of state variables, x_t, into a dynamical system of independent state variables, y_t, whose evolution can be characterized based on the analysis of the one-dimensional case examined in Sect. 1.1.

In particular, there exists a nonsingular n x n matrix, Q, whose columns are the eigenvectors of the matrix A, $\{f_1, f_2, \cdots, f_n\}$, such that

$$x_t = Qy_t + \overline{x}, \tag{3.2}$$

where $\overline{x} = [I - A]^{-1}B$ is the steady-state equilibrium of the system. Moreover,

$$y_{t+1} = Dy_t, \tag{3.3}$$

where the matrix D is a diagonal matrix whose diagonal elements are the eigenvalues of the matrix A.

$$D = \begin{bmatrix} \lambda_1 & 0 & \cdots & 0 \\ 0 & \lambda_2 & \cdots & 0 \\ \vdots & \vdots & \ddots & \vdots \\ 0 & 0 & & \lambda_n \end{bmatrix}. \tag{3.4}$$

Following the method of iterations $y_t = D^t y_0$, i.e.

$$\begin{bmatrix} y_{1t} \\ y_{2t} \\ \vdots \\ y_{nt} \end{bmatrix} = \begin{bmatrix} \lambda_1^t & 0 & \cdots & 0 \\ 0 & \lambda_2^t & \cdots & 0 \\ \vdots & \vdots & \ddots & \vdots \\ 0 & 0 & & \lambda_n^t \end{bmatrix} \begin{bmatrix} y_{10} \\ y_{20} \\ \vdots \\ y_{n0} \end{bmatrix}, \tag{3.5}$$

and the evolution of each element of the state variables y_{it} is therefore governed by

$$y_{it} = \lambda_i^t y_{i0} \quad \text{for} \quad i = 1, 2, ...n. \tag{3.6}$$

Since the vector of state variables, x_t, can be expressed as a function of y_t, namely, $x_t = Qy_t + \overline{x}$, it follows that the evolution of the vector of state variables, x_t, is

$$
\begin{bmatrix} x_{1t} \\ x_{2t} \\ \vdots \\ x_{nt} \end{bmatrix}
=
\begin{bmatrix}
Q_{11} & Q_{12} & \cdots & Q_{1n} \\
Q_{21} & Q_{22} & \cdots & Q_{2n} \\
\vdots & \vdots & \ddots & \vdots \\
Q_{n1} & Q_{n2} & \cdots & Q_{nn}
\end{bmatrix}
\begin{bmatrix} \lambda_1^t y_{10} \\ \lambda_2^t y_{20} \\ \vdots \\ \lambda_n^t y_{n0} \end{bmatrix}
+
\begin{bmatrix} \overline{x}_1 \\ \overline{x}_2 \\ \vdots \\ \overline{x}_n \end{bmatrix} . \tag{3.7}
$$

Hence,

$$x_{it} = \sum_{j=1}^{n} Q_{ij} y_{j0} \lambda_j^t + \overline{x}_i, \quad \text{for} \quad i = 1, 2, \cdots, n, \tag{3.8}$$

where

$$
\begin{bmatrix} y_{10} \\ y_{20} \\ \vdots \\ y_{n0} \end{bmatrix}
=
\begin{bmatrix}
Q_{11} & Q_{12} & \cdots & Q_{1n} \\
Q_{21} & Q_{22} & \cdots & Q_{2n} \\
\vdots & \vdots & \ddots & \vdots \\
Q_{n1} & Q_{n2} & \cdots & Q_{nn}
\end{bmatrix}^{-1}
\begin{bmatrix} x_{10} - \overline{x}_1 \\ x_{20} - \overline{x}_2 \\ \vdots \\ x_{n0} - \overline{x}_n \end{bmatrix} . \tag{3.9}
$$

Equations (3.8) and (3.9) provide the general solution for the evolution of the ith state variable, x_{it}, in terms of the eigenvalues of the matrix A, $\{\lambda_1, \lambda_2, \cdots \lambda_n\}$, the initial conditions of the entire vector of state variables, $\{x_{10}, x_{20}, \cdots x_{n0}\}$, and the ith state variable's steady-state equilibrium, $\overline{x}_i$. It sets the stage for the stability results stated in the following theorem.

Theorem 3.1. *(Necessary and Sufficient Conditions for Global Stability: Distinct Real Eigenvalues)*
Consider the system $x_{t+1} = Ax_t + B$, *where* $x_t \in \Re^n$ *and* x_0 *is given. Suppose that* $|I - A| \neq 0$ *and* A *has* n *distinct real eigenvalues* $\{\lambda_1, \lambda_2, \cdots, \lambda_n\}$.

- *a. The steady-state equilibrium* $\overline{x} = [I - A]^{-1}B$ *is globally stable if and only if*

$$|\lambda_j| < 1, \quad \forall j = 1, 2, \cdots, n.$$

  b. $\lim_{t \to \infty} x_t = \overline{x}$ *if and only if* $\forall j = 1, 2, \cdots n$

$$\{|\lambda_j| < 1 \ or \ y_{j0} = 0\},$$

  where $y_0 = Q^{-1}(x_0 - \overline{x})$, *and* Q *is a nonsingular* $n \times n$ *matrix whose columns are the eigenvectors,* $\{f_1, f_2, \cdots, f_n\}$, *of the matrix* A.

Proof:
(a) The steady-state equilibrium, $\bar{x}$, is globally stable if for all $x_0 \in \Re^n$ $\lim_{t \to \infty} x_{it} = \overline{x}_i$ for all $i = 1, 2, \cdots, n$. Thus it follows from (3.8) that global stability is satisfied if and only if $\forall K_{ij} \equiv Q_{ij}y_{j0} \in \Re$ $\lim_{t \to \infty} \sum_j K_{ij}\lambda_j^t = 0$. Namely, if and only if $|\lambda_j| < 1$, $\forall j = 1, 2, \cdots, n$,
(b) As follows from (3.8), $\lim_{t \to \infty} x_{it} = \overline{x}_i$ if and only if either $|\lambda_j| < 1$ or $[|\lambda_j| \geq 1$ and $y_{j0} = 0]$, $\forall j = 1, 2, \cdots n$. $\qquad \square$

3.1.2 Phase Diagrams of Two-Dimensional Uncoupled Systems

This subsection presents the various types of phase diagrams that characterizes the evolution of the vector of new state variables, y_t, in the two-dimensional case when the original system is characterized by a matrix of coefficients, A, that has distinct real eigenvalues. In particular, $y_{t+1} = Dy_t$, where D is a diagonal matrix with the eigenvalues of the matrix A, λ_1 and λ_2, as its diagonal elements.

As follows from the solution for the evolution of y_t given in (3.5),

$$y_{1t} = \lambda_1^t y_{10}$$

$$y_{2t} = \lambda_2^t y_{20},$$

$$(3.10)$$

and the steady-state equilibrium of the system is

$$\bar{y} = (\bar{y}_1, \bar{y}_2) = (0, 0). \tag{3.11}$$

Hence,

$$\lim_{t \to \infty} y_t = \bar{y} \quad if \quad \{|\lambda_j| < 1\} \text{ or } \{y_{j0} = 0\} \text{ for all } j = 1, 2. \tag{3.12}$$

The phase diagrams of the dynamical systems that are governed by distinct real eigenvalues depend upon the sign of each of the eigenvalues, their absolute value relative to unity, and their relative magnitude in comparison to each other.

(a) Positive Eigenvalues

If both eigenvalues are positive, then the evolution of the two state variables is necessarily monotonic. As long as the eigenvalues differ from 1, the system may converge monotonically to its steady-state equilibrium or diverge monotonically to $+\infty$ or $-\infty$ depending on the size of the eigenvalues relative to one.

- Stable Node: $0 < \lambda_2 < \lambda_1 < 1$

If both eigenvalues are positive and smaller than one, then the steady-state equilibrium, $\bar{y} = (\bar{y}_1, \bar{y}_2) = (0, 0)$, is globally stable. Namely, $\lim_{t \to \infty} y_{1t} = 0$ and $\lim_{t \to \infty} y_{2t} = 0$, $\forall (y_{10}, y_{20}) \in \Re^2$.

As depicted in Fig. 3.1, the state variables monotonically converge to the steady-state equilibrium, $(0, 0)$, as time approaches infinity. However, since $\lambda_2 < \lambda_1$ the convergence of y_{2t} to 0 is faster than the convergence of y_{1t} to 0.

- Saddle: $0 < \lambda_2 < 1 < \lambda_1$

The steady-state equilibrium is a saddle. Namely, $\lim_{t \to \infty} y_{2t} = 0 \ \forall y_{20} \in \Re$, whereas $\lim_{t \to \infty} y_{1t} = 0$ if and only if $y_{10} = 0$. As depicted in Fig. 3.2, the convergence to the steady-state equilibrium $\bar{y}$ along the saddle path (i.e. along the stable eigenspace, or alternatively, the stable manifold) is monotonic, and the diverging paths are monotonic as well.

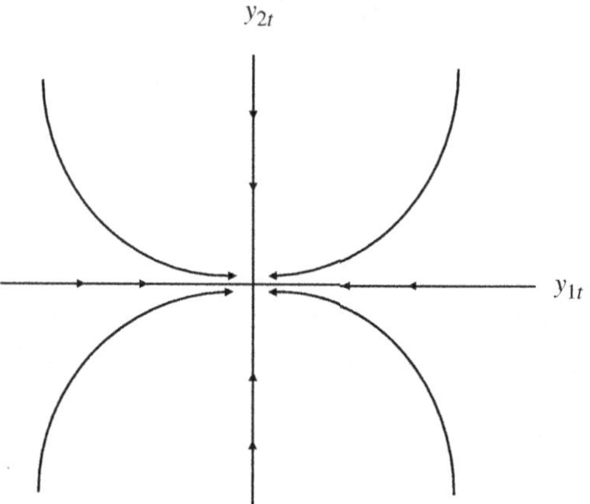

Fig. 3.1. Stable Node
$$0 < \lambda_2 < \lambda_1 < 1$$

- Focus: $0 < \lambda_1 = \lambda_2 < 1$

The steady-state equilibrium is globally stable. Namely, $\lim_{t\to\infty} y_{1t} = 0$ and $\lim_{t\to\infty} y_{2t} = 0$, $\forall(y_{10}, y_{20}) \in \Re^2$. As depicted in Fig. 3.3 convergence to the steady-state equilibrium, $(0,0)$, is monotonic. The speed of convergence is the same for each state variable, and the trajectory of the system from any initial condition to the steady-state equilibrium is linear.

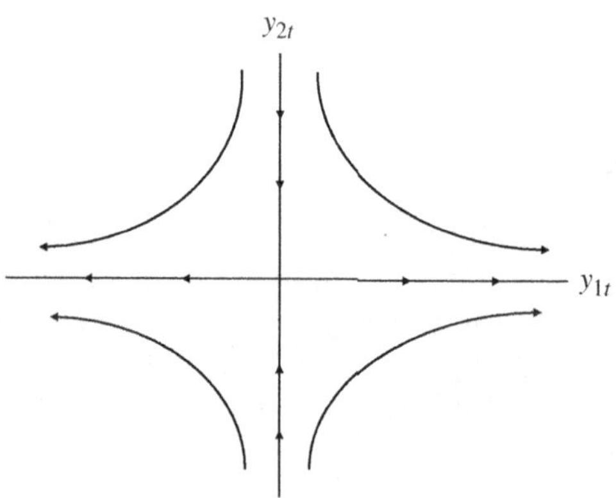

Fig. 3.2. A Saddle
$$0 < \lambda_2 < 1 < \lambda_1$$

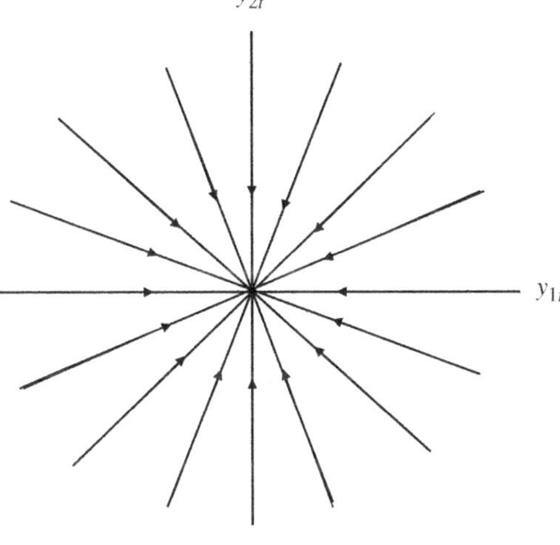

Fig. 3.3. Focus
$$0 < \lambda_1 = \lambda_2 < 1$$

- Source: $1 < \lambda_1 < \lambda_2$

The steady-state equilibrium, $(0,0)$, is unstable. Namely, $\lim_{t \to \infty} y_{1t} = \pm\infty$ and $\lim_{t \to \infty} y_{2t} = \pm\infty$, $\forall(y_{10}, y_{20}) \in \Re^2 - \{0\}$ (i.e., as long as the initial conditions are different from zero). As depicted in Fig. 3.4, the divergence of the two state variables is monotonic. However, since $\lambda_2 > \lambda_1$ the divergence of y_{2t} is faster than the divergence of y_{1t}.

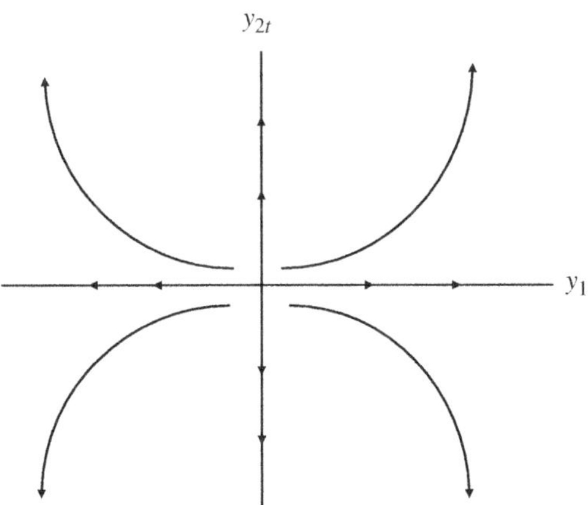

Fig. 3.4. A Source
$$1 < \lambda_1 < \lambda_2.$$

(b) Negative Eigenvalues

The evolution of each of the two state variables is determined by their associated eigenvalues. If one of the eigenvalues is positive and the other is negative, then the variable associated with the positive eigenvalue converges or diverges monotonically while the variable associated with the negative eigenvalue is characterized by oscillatory convergence or divergence. The evolution of the two state variables is therefore reflected around one of the axes. However, if both eigenvalues are negative, each of the state variables evolves in oscillations between negative and positive values. The evolution of the system is therefore reflected around two axes. In both cases the system, may converge to its steady-state level, or diverge, depending on the absolute value of the eigenvalues relative to one.

- Stable Node (oscillatory convergence of one state variable): $[-1 < \lambda_i < 0 < \lambda_j < 1, \ \text{for} \ i,j = 1,2]$

Since the absolute value of both eigenvalues is smaller than one, the system converges globally towards the steady-state equilibrium $\overline{y} = (\overline{y}_1, \overline{y}_2) = (0,0)$.

As depicted in Fig. 3.5, y_{1t} converges monotonically whereas y_{2t} converges in oscillations, reflected around the y_{1t} axis. Additionally, if

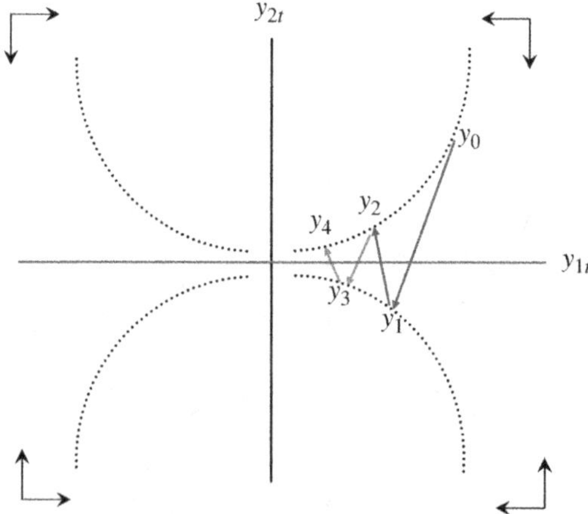

Fig. 3.5. Stable Node: Oscillatory Convergence of one State Variable
$$-1 < \lambda_2 < 0 < \lambda_1 < 1$$

$|\lambda_1| > |\lambda_2|$, the convergence of y_{2t} is faster towards its steady-state level.

- Stable Node (oscillatory convergence of both state variables)
  $[-1 < \lambda_i < \lambda_j < 0$ for $i, j = 1, 2]$

The steady-state equilibrium is globally stable. As depicted in Fig. 3.6, the convergence of both variables towards the steady-state equilibrium is oscillatory. Moreover, if $|\lambda_1| > |\lambda_2|$ the convergence of y_{2t} is faster towards its steady-state level.

- Saddle (oscillatory convergence and divergence)$[\{\lambda_i < -1 < \lambda_j < 0\}$, $\{\lambda_i < -1$ and $0 < \lambda_j < 1\}$, or $\{\lambda_i > 1$ and $-1 < \lambda_j < 0\}$ for $i, j = 1, 2]$

If both eigenvalues are negative, one variable converges in an oscillatory manner while the other diverges in an oscillatory manner, whereas if one is negative and one is positive, one variable converges monotonically while the other diverges in an oscillatory manner. Figure 3.7 depicts the dynamical system for the case in which $\lambda_1 > 1$ and $-1 < \lambda_2 < 0$. Convergence to the steady-state equilibrium $\overline{y} = (\overline{y}_1, \overline{y}_2) = (0, 0)$ occurs in oscillations along the vertical axis, and divergence in oscillations takes place as long as $y_{1t} \neq 0$.

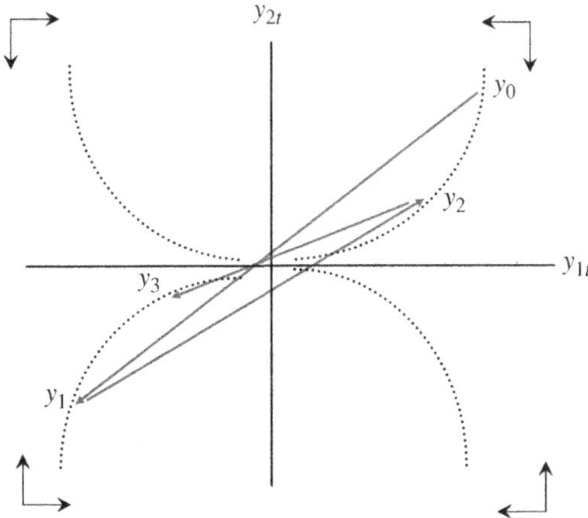

Fig. 3.6. Stable Node: Oscillatory Convergence of Both State Variables
$-1 < \lambda_1 < \lambda_2 < 0$

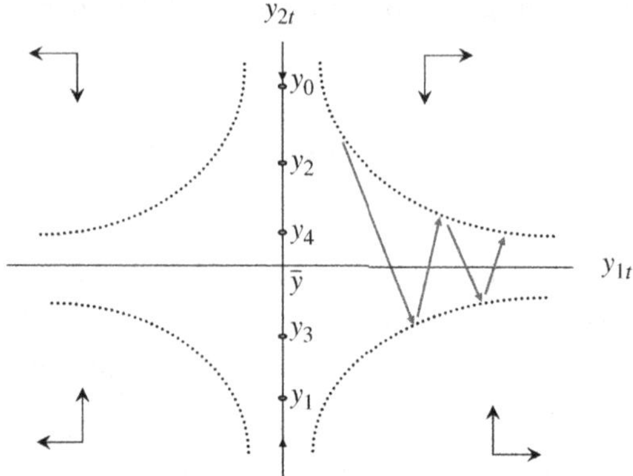

Fig. 3.7. Saddle: Oscillatory Convergence and Divergence
$-1 < \lambda_2 < 0$ and $\lambda_1 > 1$

- Focus (oscillatory convergence): $-1 < \lambda_1 = \lambda_2 < 0$

The steady-state equilibrium is globally stable. Convergence of both state variables is oscillatory.

- Source (oscillatory divergence): $[\{\lambda_i < \lambda_j < -1\}$ or $\{\lambda_i < -1$ and $\lambda_j > 1\}$ for $i, j = 1, 2]$

The steady-state equilibrium is unstable. Divergence of negative eigenvalues is oscillatory whereas divergence of positive eigenvalues is monotonic.

3.2 Repeated Real Eigenvalues

3.2.1 Characterization of the Solution

Consider the multi-dimensional, linear dynamical system

$$x_{t+1} = Ax_t + B, \tag{3.13}$$

where the n x n matrix of coefficients, A, has n repeated real eigenvalues, $\{\lambda, \lambda, \cdots, \lambda\}$. Suppose further that $|I - A| \neq 0$.

As was established in Lemma 2.6 and (2.88), there exists a time-invariant transformation of the vector of state variables x_t into a dynamical system of interdependent state variables y_t, whose evolution can be characterized based on the properties of the Jordan matrix.

In particular, there exists a nonsingular $n \times n$ matrix, Q, such that,

$$x_t = Qy_t + \overline{x}, \qquad (3.14)$$

where $\overline{x} = [I - A]^{-1}B$ is the steady-state equilibrium of the system. Moreover,

$$y_{t+1} = Dy_t, \qquad (3.15)$$

where the matrix D is

$$D = \begin{bmatrix} \lambda & 0 & 0 & 0 & \cdots & 0 \\ 1 & \lambda & 0 & 0 & \ddots & 0 \\ 0 & 1 & \lambda & 0 & \ddots & 0 \\ 0 & 0 & 1 & \ddots & \ddots & 0 \\ 0 & 0 & 0 & \ddots & \lambda & 0 \\ 0 & 0 & 0 & \cdots & 1 & \lambda \end{bmatrix}. \qquad (3.16)$$

Hence, the evolution of the first new state variable y_{1t} is independent of all other state variables except the first, the evolution of the second state variable depends on the evolution of the first and the second state variable, the evolution of the third state variable depends directly only on the second and the third, and so forth.

The evolution of the vector of state variables y_t is therefore given by

$$y_t = D^t y_0, \qquad (3.17)$$

where for $t > n$,

$$D^t = \begin{bmatrix} \lambda^t & 0 & 0 & \cdots & 0 \\ t\lambda^{t-1} & \lambda^t & 0 & \ddots & 0 \\ \dfrac{t(t-1)\lambda^{t-2}}{2!} & t\lambda^{t-1} & \lambda^t & \ddots & 0 \\ \vdots & & & \ddots & 0 \\ \dfrac{t(t-1)\cdots(t-n+2)\lambda^{t-n+1}}{(n-1)!} & & \cdots & t\lambda^{t-1} & \lambda^t \end{bmatrix}.$$

$$\qquad (3.18)$$

Hence, the value of the vector of state variables y_t for $t > n$ is

$$y_{1t} = \lambda^t y_{10}$$

$$y_{2t} = t\lambda^{t-1} y_{10} + \lambda^t y_{20}$$

$$y_{3t} = \frac{t(t-1)\lambda^{t-2}}{2!} y_{10} + t\lambda^{t-1} y_{20} + \lambda^t y_{30} \qquad (3.19)$$

$$\vdots \qquad \vdots$$

$$y_{nt} = \frac{t(t-1)\cdots(t-n+2)\lambda^{t-n+1}}{(n-1)!} y_{10} + \cdots + \lambda^t y_{n0}.$$

For all $i = 1, 2, \cdots, n$, it therefore follows that

$$y_{it} = \sum_{k=0}^{i-1} \binom{t}{k} \lambda^{t-k} y_{i-k,0}, \qquad (3.20)$$

where

$$\binom{t}{k} = \frac{t!}{k!(t-k)!}. \qquad (3.21)$$

Since the vector of state variables x_t can be expressed as a function of y_t, namely, $x_t = Qy_t + \bar{x}$, it follows that the evolution of the vector of state variables x_t is

$$\begin{bmatrix} x_{1t} \\ x_{2t} \\ \vdots \\ x_{nt} \end{bmatrix} = \begin{bmatrix} Q_{11} & Q_{12} & \cdots & Q_{1n} \\ Q_{21} & Q_{22} & \cdots & Q_{2n} \\ \vdots & \vdots & & \vdots \\ Q_{n1} & Q_{n2} & \cdots & Q_{nn} \end{bmatrix} \begin{bmatrix} \lambda_1^t y_{10} \\ t\lambda^{t-1} y_{10} + \lambda^t y_{20} \\ \vdots \\ \frac{t(t-1)\cdots(t-n+2)\lambda^{t-n+1}}{(n-1)!} y_{10} + \cdots + \lambda^t y_{n0} \end{bmatrix}$$

$$+ \begin{bmatrix} \bar{x}_1 \\ \bar{x}_2 \\ \vdots \\ \bar{x}_n \end{bmatrix}. \qquad (3.22)$$

Hence, $\forall i = 1, 2, \cdots, n$,

$$x_{it} = \sum_{m=0}^{n-1} \binom{t}{m} \lambda^{t-m} K_{i,m+1} + \bar{x}_i, \qquad (3.23)$$

where $K_{i,m+1}$ are constants that reflect the product of the ith row of matrix Q and the initial conditions $(y_{10}, y_{20}, \cdots, y_{n0})$.

Hence the evolution of the vector of the original state variables, x_t, is governed by the value of the repeated eigenvalue, λ, and the initial conditions. This solution sets the stage for the stability result stated in the following theorem.

Theorem 3.2. *(Necessary and Sufficient Conditions for Global Stability: Repeated Real Eigenvalues)*
Consider the system $x_{t+1} = Ax_t + B$, where $x_t \in \Re^n$ and x_0 is given. Suppose that $|I - A| \neq 0$ and A has n repeated real eigenvalues $\{\lambda, \lambda, \cdots, \lambda\}$. Then, the steady-state equilibrium, $\bar{x} = [I - A]^{-1}B$, is globally stable if and only if

$$|\lambda| < 1.$$

Proof: Follows from (3.23). $\qquad\qquad\qquad\qquad\qquad\qquad\qquad\qquad$ $\square$

It should be noted that if $|\lambda| \geq 1$ the system does not converge to its steady-state equilibrium unless it starts at this equilibrium point. If $|\lambda| \geq 1$ then $\lim_{t \to \infty} x_t = \bar{x}$ only if $y_{i0} = 0$, $\forall i = 1, 2, \cdots, n$, and thus, since $y_0 = Q^{-1}(x_0 - \bar{x})$, only if $x_{i0} = \bar{x}_i$, $\forall i = 1, 2, \cdots, n$.

3.2.2 Phase Diagram of the Two-Dimensional Case

This subsection presents the phase diagrams that characterize the evolution of the vector of new state variables, y_t, in the two-dimensional case, when the original system is characterized by a matrix of coefficients, A, that has repeated real eigenvalues.

The evolution of the vector of the new state variables, y_t, from time t to time $t + 1$, is given by

$$\begin{bmatrix} y_{1t+1} \\ y_{2t+1} \end{bmatrix} = \begin{bmatrix} \lambda & 0 \\ 1 & \lambda \end{bmatrix} \begin{bmatrix} y_{1t} \\ y_{2t} \end{bmatrix}. \qquad (3.24)$$

Hence, following the method of iterations,

$$y_t = D^t y_0, \qquad (3.25)$$

where

$$D^t = \begin{bmatrix} \lambda^t & 0 \\ t\lambda^{t-1} & \lambda^t \end{bmatrix}. \qquad (3.26)$$

The trajectory of the two state variables is given therefore by

$$y_{1t} = \lambda^t y_{10}$$
$$y_{2t} = t\lambda^{t-1} y_{10} + \lambda^t y_{20}. \tag{3.27}$$

The derivation of the phase diagram of this uncoupled system is simpler in the context of the derivation of the forces that operate on the system when it is not in a steady-state equilibrium . The system takes the form of

$$y_{1t+1} = \lambda y_{1t}$$
$$y_{2t+1} = y_{1t} + \lambda y_{2t}, \tag{3.28}$$

and therefore the changes in the value of each of the new state variables from period t to period $t + 1$, Δy_{it}, $i = 1, 2$, is

$$\Delta y_{1t} \equiv y_{1t+1} - y_{1t} = -(1 - \lambda)y_{1t}$$
$$\Delta y_{2t} \equiv y_{2t+1} - y_{2t} = y_{1t} - (1 - \lambda)y_{2t}. \tag{3.29}$$

Consequently,

$$\Delta y_{1t} = 0 \Leftrightarrow \{y_{1t} = 0 \text{ or } \lambda = 1\}$$
$$\Delta y_{2t} = 0 \Leftrightarrow \{(y_{2t} = \tfrac{y_{1t}}{1-\lambda} \text{ and } \lambda \neq 1) \text{ or } (y_{1t} = 0 \text{ and } \lambda = 1)\}. \tag{3.30}$$

The phase diagram of the dynamical system depends upon the absolute magnitude of the eigenvalue relative to unity and on its sign.

(a) Positive Eigenvalues

If the repeated eigenvalue is positive, then the evolution of the two state variables is not oscillatory, but nevertheless it is not necessarily monotonic. As long as the eigenvalues differ from 1, the system may converge to its steady-state equilibrium, or diverge monotonically to $\pm\infty$ depending upon the size of the eigenvalues relative to one.

- **Improper (Stable) Node:** $\lambda \in (0, 1)$

Since the repeated eigenvalue is smaller than 1, the steady-state equilibrium, $(0, 0)$, is globally stable. However, although the eigenvalue is positive, convergence to the steady-state equilibrium is not monotonic.

As follows from (3.30)

$$\Delta y_{1t} = 0 \ \Leftrightarrow \ y_{1t} = 0. \tag{3.31}$$

Namely, as depicted in Fig. 3.8, the y_{2t} axis is the geometric place of all pairs (y_{1t}, y_{2t}) such that $\Delta y_{1t} = 0$. Furthermore,

$$\Delta y_{1t} \begin{cases} > 0 & \text{if} \quad y_{1t} < 0 \\ \\ < 0 & \text{if} \quad y_{1t} > 0 \end{cases}. \tag{3.32}$$

Similarly,

$$\Delta y_{2t} = 0 \ \Leftrightarrow \ y_{2t} = \frac{y_{1t}}{(1-\lambda)}. \tag{3.33}$$

Namely, the '$\Delta y_{2t} = 0$' locus is a line with a slope greater than unity and

$$\Delta y_{2t} \begin{cases} < 0 & \text{if} \quad y_{2t} > \frac{y_{1t}}{1-\lambda} \\ \\ > 0 & \text{if} \quad y_{2t} < \frac{y_{1t}}{1-\lambda}. \end{cases} \tag{3.34}$$

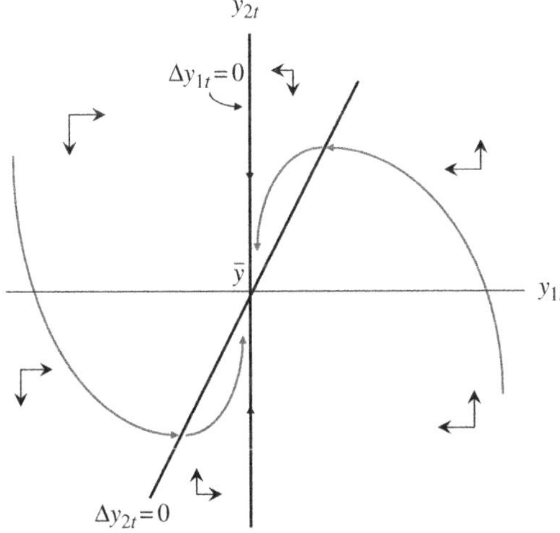

Fig. 3.8. Improper Stable Node
$\lambda \in (0, 1)$

As follows directly from (3.27), noting that $0 < \lambda < 1$, or indirectly from (3.32), the state variable y_{1t} converges monotonically to the steady-state equilibrium, $\bar{y} = 0$. In contrast, the second state variable y_{2t} converges to the steady-state in a non-monotonic fashion.

As depicted in Figs. 3.8 and 3.9, if $y_{2t} < 0$ and $y_{1t} > 0$, then y_{2t} increases monotonically, crossing to the positive quadrant and peaking when it meets the '$\Delta y_{2t} = 0$' locus. Afterwards it decreases monotonically and converges to the steady-state equilibrium $\bar{y} = 0$.

Remark. The trajectories in Fig. 3.8 are drawn based on additional necessary information. In particular, it should be noted that if the system is in quadrants I or IV it cannot cross into quadrants II or III, and vice versa. This is the case since, as follows from (3.24), if $y_{1t} > 0$ then $y_{1t+1} > 0$ whereas if $y_{1t} < 0$ then $y_{1t+1} < 0$. The system, therefore never crosses the y_2 - axis. Furthermore, as follows from (3.24), if the system enters quadrant I or III, it never leaves them. This is the case since if $y_{1t} > 0$ and $y_{2t} > 0$ then $y_{1t+1} > 0$ and $y_{2t+1} > 0$ whereas if $y_{1t} < 0$ and $y_{2t} < 0$ then $y_{1t+1} < 0$ and $y_{2t+1} < 0$.

- **Improper Source**: $\lambda \in (1, \infty)$

Since the repeated eigenvalue is larger than 1, the steady-state equilibrium is globally unstable. However, although the eigenvalue is positive, divergence is not monotonic.

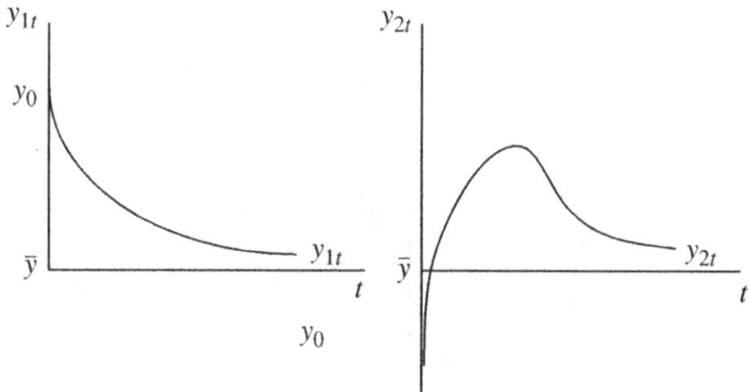

Fig. 3.9. Monotonic Evolution of y_{1t} and Non-Monotonic evolution of y_{2t}
$\lambda \in (0, 1)$

As follows from (3.29) and (3.30),

$$
\Delta y_{1t}
\begin{cases}
> 0 & \Leftrightarrow \quad y_{1t} > 0 \\
= 0 & \Leftrightarrow \quad y_{1t} = 0 \;, \\
< 0 & \Leftrightarrow \quad y_{1t} < 0
\end{cases}
\tag{3.35}
$$

whereas

$$
\Delta y_{2t}
\begin{cases}
> 0 & \Leftrightarrow \quad y_{2t} > \frac{y_{1t}}{1-\lambda} \\
= 0 & \Leftrightarrow \quad y_{2t} = \frac{y_{1t}}{1-\lambda} \;. \\
< 0 & \Leftrightarrow \quad y_{2t} < \frac{y_{1t}}{1-\lambda}
\end{cases}
\tag{3.36}
$$

Hence, as depicted in Fig. 3.10, the geometric locus '$\Delta y_{1t} = 0$' coincides with the y_{2t} axis, as was the case when $\lambda \in (0,1)$, whereas the geometric locus '$\Delta y_{2t} = 0$' is a line with negative slope $1/(1-\lambda)$.

As follows directly from (3.27), noting that $\lambda > 1$, or indirectly from (3.32), the state variable y_{1t} diverges monotonically to either $+\infty$ or $-\infty$. In contrast, the second state variable, y_{2t}, diverges non-montonically to either $+\infty$ or $-\infty$. As depicted in Fig. 3.10, if $y_{2t} < 0$ and $y_{1t} > 0$, then y_{2t} decreases monotonically and starts increasing indefinitely when it crosses the '$\Delta y_{2t} = 0$' locus.

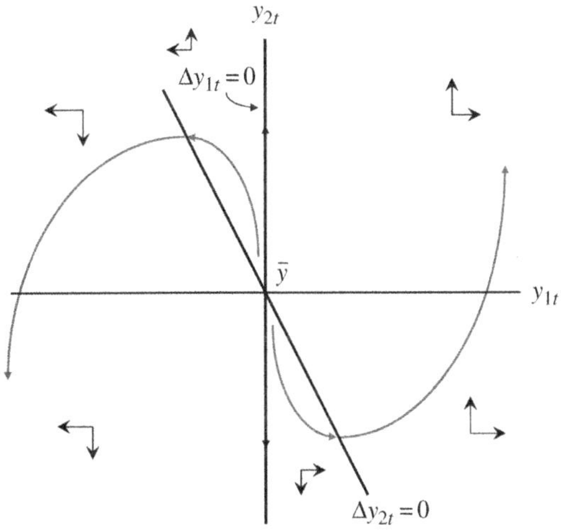

Fig. 3.10. Improper Source
$\lambda \in (1, \infty)$

(b) Negative Eigenvalues

The evolution of the two state variables is characterized by oscillations reflected around the two axes. The state variables may converge to their steady-state level, or diverge, depending on the absolute value of the eigenvalues relative to one.

- **Improper Stable Node:** $\lambda \in (-1, 0)$ (Oscillatory Convergence)

  If the repeated eigenvalue is negative but smaller than one in absolute value, then the system converges to the steady-state equilibrium, $(0, 0)$. Depending on the initial conditions, the system oscillates between either quadrants IV and II or I and III.

- **Improper Source:** $\lambda \in (-\infty, 1)$ (Oscillatory Divergence)

  If the repeated eigenvalue is negative but greater than one in absolute value, then the system diverges to $+\infty$ or $-\infty$. Depending on the system's initial conditions, the system oscillates between either quadrants IV and II or I and III.

(c) Eigenvalue of Modulus One

If the repeated eigenvalue is in absolute value equal to 1, then the system is characterized by a continuum of unstable equilibria. This non-generic case represents the bifurcation point of the dynamical system. Namely, an infinitesimal change in the value of λ brings about a *qualitative* change in the nature of the dynamical system. In particular, if λ declines continuously from a value greater than 1 to a value below 1, the set of steady-state equilibria changes from a unique unstable steady-state equilibrium to a continuum of unstable equilibria and then to a unique globally stable steady-state equilibrium.

- **Continuum of Unstable Steady-State Equilibria:** $\lambda = 1$

  As follows from (3.29) and (3.30),

$$\Delta y_{1t} = 0 \quad \text{for all } (x_{it}, x_{2t}) \in \Re^2, \tag{3.37}$$

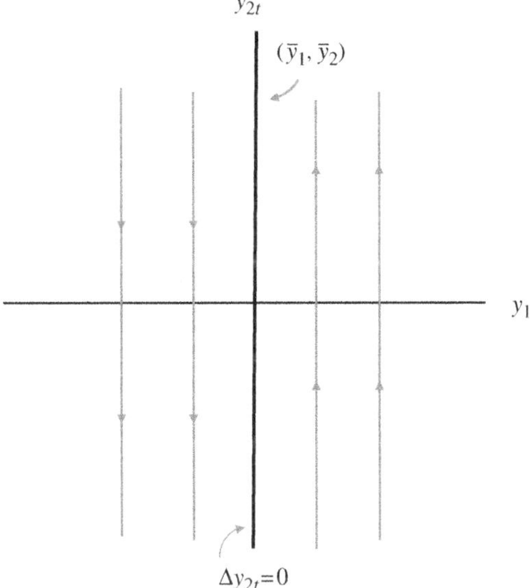

Fig. 3.11. Continuum of Unstable Steady-State Equilibria
$$\lambda = 1$$

whereas

$$\Delta y_{2t} \begin{cases} > 0 & \Leftrightarrow \quad y_{1t} > 0 \\[2mm] = 0 & \Leftrightarrow \quad y_{1t} = 0 \\[2mm] < 0 & \Leftrightarrow \quad y_{1t} < 0. \end{cases} \qquad (3.38)$$

Thus the geometric place under which the first state variable, y_{1t}, is in a steady-state equilibrium is the entire $\Re^2$ plane, whereas the geometric place under which the second state variable, y_{2t}, is in steady-state equilibrium is the y_{2t} axis. The set of steady-state equilibria for the system is therefore the entire y_{2t} axis. However, as depicted in Fig. 3.11, none of these equilibria is stable.

3.3 Distinct Pairs of Complex Eigenvalues

3.3.1 Characterization of the Solution

Consider the multi-dimensional, linear dynamical system

$$x_{t+1} = Ax_t + B. \qquad (3.39)$$

Suppose that the n x n matrix of coefficients, A, has $n/2$ pairs of distinct complex eigenvalues $\{\mu_1, \overline{\mu}_1, \mu_2, \overline{\mu}_2, \cdots, \mu_{n/2}, \overline{\mu}_{n/2}\}$, where

$$\mu_j \equiv \alpha_j + \beta_j i,$$

$$\overline{\mu}_j \equiv \alpha_j - \beta_j i, \tag{3.40}$$

and $i \equiv \sqrt{-1}$. Suppose further that $|I - A| \neq 0$.[1]

As was established in Lemma 2.7 and (2.88), there exists a time-invariant transformation of the vector of state variables, x_t, into a dynamical system of interdependent state variables, y_t, whose evolution can be characterized based on the properties of a matrix in the *Jordan normal form*.

In particular, there exists a nonsingular n x n matrix, Q, such that

$$x_t = Q y_t + \overline{x}, \tag{3.41}$$

where $\overline{x} = [I - A]^{-1} B$ is the steady-state equilibrium of the system, and

$$y_{t+1} = D y_t, \tag{3.42}$$

where

$$
D = \begin{bmatrix}
\alpha_1 & -\beta_1 & 0 & 0 & \ldots & \ldots & 0 & 0 \\
\beta_1 & \alpha_1 & 0 & 0 & \ldots & \ldots & 0 & 0 \\
0 & 0 & \alpha_2 & -\beta_2 & \ldots & \ldots & 0 & 0 \\
0 & 0 & \beta_2 & \alpha_2 & \ldots & \ldots & 0 & 0 \\
0 & 0 & 0 & 0 & \ddots & \ddots & 0 & 0 \\
0 & 0 & 0 & 0 & \ddots & \ddots & 0 & 0 \\
0 & 0 & 0 & 0 & \ldots & \ldots & \alpha_{n/2} & -\beta_{n/2} \\
0 & 0 & 0 & 0 & \ldots & \ldots & \beta_{n/2} & \alpha_{n/2}
\end{bmatrix}. \tag{3.43}
$$

---

[1] If n were odd, then the additional eigenvalue would necessarily be real.

Thus, each pair of state variables, $\{y_{2j-1,t}, y_{2j,t}\}$, evolves independently of all other pairs. In particular, for all $j = 1, 2, \cdots, n/2$,

$$\begin{bmatrix} y_{2j-1,t+1} \\ y_{2j,t+1} \end{bmatrix} = \begin{bmatrix} \alpha_j & -\beta_j \\ \beta_j & \alpha_j \end{bmatrix} \begin{bmatrix} y_{2j-1,t} \\ y_{2j,t} \end{bmatrix}. \tag{3.44}$$

Following the method of iterations, the trajectory of the evolution of each pair of state variables $\{y_{2j-1,t}, y_{2j,t}\}_0^\infty$, $j = 1, 2, \cdots, n/2$, satisfies the equation

$$\begin{bmatrix} y_{2j-1,t} \\ y_{2j,t} \end{bmatrix} = \begin{bmatrix} \alpha_j & -\beta_j \\ \beta_j & \alpha_j \end{bmatrix}^t \begin{bmatrix} y_{2j-1,0} \\ y_{2j,0} \end{bmatrix}. \tag{3.45}$$

This formulation, however, is not very informative about the qualitative behavior of the dynamical system. In particular, it is not apparent what the necessary restrictions on the values of α_j and β_j are, such that each pair j of state variables will converge to its steady-state value.

However, the evolution of each pair of state variables can be expressed in terms of the "polar coordinates" of (α_j, β_j), and then the necessary restrictions on the values of α_j and β_j that assure the stability of the dynamical system becomes apparent.

Consider the geometrical representation of the complex pair of eigenvalues, $\mu_j \equiv \alpha_j + \beta_j i$ and $\bar{\mu}_j \equiv \alpha_j - \beta_j i$, in the complex Cartesian space as depicted in Fig. 3.12.

Let

$$r_j \equiv \sqrt{(\alpha_j^2 + \beta_j^2)}. \tag{3.46}$$

Namely, r_j is the modulus of the jth eigenvalue. It follows that

$$\alpha_j = r_j \cos \theta_j,$$

$$\beta_j = r_j \sin \theta_j, \tag{3.47}$$

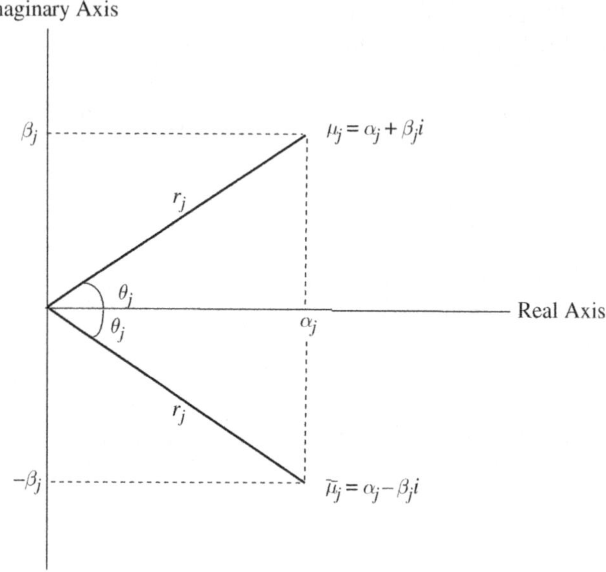

Fig. 3.12. The Complex Cartesian Space

and therefore

$$\begin{bmatrix} \alpha_j & -\beta_j \\ \beta_j & \alpha_j \end{bmatrix} = r_j \begin{bmatrix} \cos\theta_j & -\sin\theta_j \\ \sin\theta_j & \cos\theta_j \end{bmatrix}. \tag{3.48}$$

Hence, as follows from (3.45) and (3.48), the evolution of each pair of state variables $\{y_{2j-1,t}, y_{2j,t}\}$ can be determined by well-established trigonometric properties.

Lemma 3.3.

$$\left\{ r_j \begin{bmatrix} \cos\theta_j & -\sin\theta_j \\ \sin\theta_j & \cos\theta_j \end{bmatrix} \right\}^t = r_j^t \begin{bmatrix} \cos t\theta_j & -\sin t\theta_j \\ \sin t\theta_j & \cos t\theta_j \end{bmatrix}.$$

Proof: The lemma follows from the trigonometric identities:

$$\begin{aligned} \cos(\theta_1 + \theta_2) &= \cos\theta_1 \cos\theta_2 - \sin\theta_1 \sin\theta_2 \\ \sin(\theta_1 + \theta_2) &= \cos\theta_1 \sin\theta_2 + \sin\theta_1 \cos\theta_2. \end{aligned} \tag{3.49}$$

$\square$

The evolution of each pair of state variables $\{y_{2j-1,t}, y_{2j,t}\}$ is therefore given by

$$\begin{bmatrix} y_{2j-1,t} \\ y_{2j,t} \end{bmatrix} = r_j^t \begin{bmatrix} \cot s\theta_j & -\sin t\theta_j \\ \sin t\theta_j & \cos t\theta_j \end{bmatrix} \begin{bmatrix} y_{2j-1,0} \\ y_{2j,0} \end{bmatrix}. \tag{3.50}$$

Since the vector of the original state variables, x_t, can be expressed as a function of y_t, namely, $x_t = Qy_t + \overline{x}$, it follows that the evolution of the vector of state variables x_t is

$$\begin{bmatrix} x_{1t} \\ x_{2t} \\ \vdots \\ x_{nt} \end{bmatrix} = \begin{bmatrix} Q_{11} & Q_{12} & \cdots & Q_{1n} \\ Q_{21} & Q_{22} & \cdots & Q_{2n} \\ \vdots & \vdots & & \vdots \\ Q_{n1} & Q_{n2} & & Q_{nn} \end{bmatrix} \begin{bmatrix} r_1^t(\cos t\theta_1 y_{10} - \sin t\theta_1 y_{20}) \\ r_1^t(\sin t\theta_1 y_{10} + \cos t\theta_1 y_{20}) \\ \vdots \\ \vdots \end{bmatrix} + \begin{bmatrix} \overline{x}_1 \\ \overline{x}_2 \\ \vdots \\ \overline{x}_n \end{bmatrix}. \tag{3.51}$$

The evolution of each of the original state variables, x_{it}, $i = 1, 2, ..., n$, is therefore given by

$$x_{it} = \sum_j r_j^t [K_{ij} \cos t\theta_j + \tilde{K}_{ij} \sin t\theta_j] + \overline{x}_i, \tag{3.52}$$

where $K_{ij} \equiv Q_{i,2j-1}y_{2j-1,0} + Q_{i,2j}y_{2j,0}$ and $\tilde{K}_{ij} \equiv Q_{i,2j}y_{2j-1,0} - Q_{i,2j-1}y_{2j,0}$, for all $j = 1, 2, \cdots, n/2$.

Theorem 3.4. *(Necessary and Sufficient Conditions for Global Stability: Distinct Complex Eigenvalues)*
Consider the dynamical system $x_{t+1} = Ax_t + B$, where $x_t \in \Re^n$. Suppose that $|I - A| \neq 0$ and suppose that A has $n/2$ pairs of distinct complex eigenvalues $\{\mu_1, \overline{\mu}_1, \mu_2, \overline{\mu}_2, \cdots, \mu_{n/2}, \overline{\mu}_{n/2}\}$, where $\mu_j \equiv \alpha_j + \beta_j i$, $\overline{\mu}_j = \alpha_j - \beta_j i$, and $i \equiv \sqrt{-1}$, $j = 1, 2, \cdots, n/2$. Then the steady-state equilibrium of the dynamical system, $\overline{x}$, is globally (asymptotically) stable if and only if the modulus of each eigenvalue of the matrix A is smaller than 1, i.e. if

$$r_j \equiv \sqrt{(\alpha_j^2 + \beta_j^2)} < 1, \quad \forall j = 1, 2, \cdots, n/2.$$

Proof: Since for all t, $0 \leq |\cos t\theta_j| \leq 1$ and $0 \leq |\sin t\theta_j| \leq 1$, it follows from (3.52) that $\lim_{t \to \infty} x_{it} = \overline{x}_i$ if and only if $r_j < 1$ $\forall j = 1, 2, \cdots, n/2$. $\qquad \square$

3.3.2 Phase Diagram of a Two-Dimensional System

This subsection presents the phase diagrams that characterize the evolution of the vector of new state variables, y_t, in the two-dimensional case when the original system is characterized by a matrix of coefficients, A, that has a distinct pair of complex eigenvalues.

The evolution of vector of the state variables, y_t, from time t to time $t + 1$ is given by

$$\begin{bmatrix} y_{1t+1} \\ y_{2t+1} \end{bmatrix} = \begin{bmatrix} \alpha & -\beta \\ \beta & \alpha \end{bmatrix} \begin{bmatrix} y_{1t} \\ y_{2t} \end{bmatrix}. \tag{3.53}$$

The asymptotic behavior of the dynamical system will be determined by the modulus of the eigenvalue, r, and the values of α and β. As established in Theorem 3.4, the value of r determines whether the steady-state equilibrium is globally stable or whether the system is characterized by divergence or periodic orbit. Moreover, the motion of the system is spiral, with orientation that is determined by the sign of β and pace that is determined by the value of α and β.

(a) Periodic Orbit: $r = 1$

- **Counter-Clockwise Periodic Orbit: $\beta > 0$**

The dynamical system, as depicted in Fig. 3.13(a), exhibits a counter-clockwise periodic orbit. Since $r = 1$, it follows that $(\alpha_j^2 + \beta_j^2) = 1$. For the sake of exposition, suppose that $\beta = 1$ and consequently $\alpha = 0$. Suppose further that the initial condition is $(y_{10}, y_{20}) =$

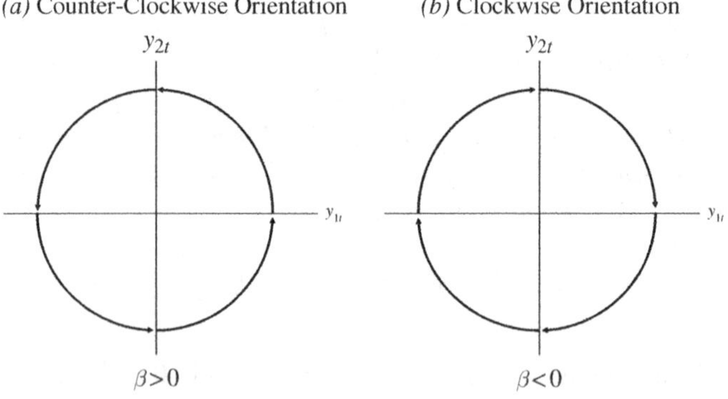

Fig. 3.13. Period Orbit

$(1, 0)$. It follows from (3.53) that $(y_{11}, y_{21}) = (0, 1)$, $(y_{12}, y_{22}) = (-1, 0)$, $(y_{13}, y_{23}) = (0, -1)$, and $(y_{14}, y_{24}) = (1, 0)$. Thus the system is characterized in this example by a four-period cycle with counter-clockwise orientation. Similarly, if $\beta = \alpha = 1/\sqrt{2}$ then $(y_{11}, y_{21}) = (1/\sqrt{2}, 1/\sqrt{2}), (y_{12}, y_{22}) = (0, 1)$, ..., and the system is characterized by an eight-period cycle with a counter-clockwise orientation.

- **Clockwise Periodic Orbit**: $\beta < 0$

The dynamical system, as depicted in Fig. 3.13(b), exhibits a clockwise periodic orbit. Since $r = 1$, it follows that $(\alpha_j^2 + \beta_j^2) = 1$. For the sake of exposition, suppose that $\beta = -1$ and consequently $\alpha = 0$. Suppose further that the initial condition is $(y_{10}, y_{20}) = (1, 0)$. The system exhibits again a four-period cycle, $\{(1, 0), (0, -1), (-1, 0), (0, 1)\}$, but with clockwise orientation.

(b) Spiral Sink: $r < 1$

The dynamical system, as depicted in Fig. 3.14, is characterized by spiral convergence towards the steady-state equilibrium, $(0, 0)$. If $\beta > 0$ the motion is counter-clockwise whereas if $\beta < 0$ the motion is clockwise.

(c) Spiral Source: $r > 1$

The system, as depicted in Fig. 3.15, exhibits spiral divergence from its steady-state equilibrium, $(0, 0)$, with either counter-clockwise motion $(\beta > 0)$ or clockwise motion $(\beta < 0)$.

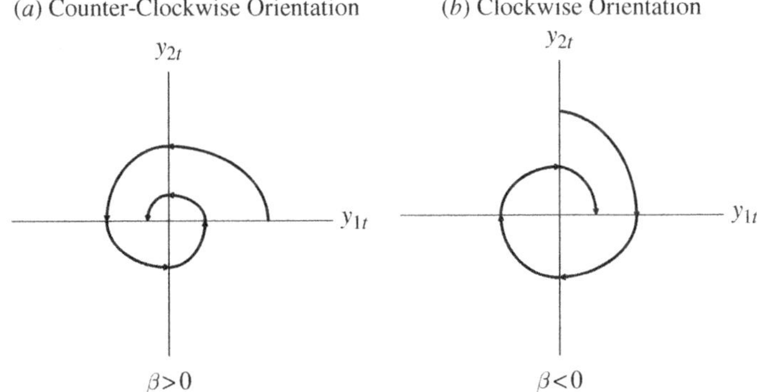

Fig. 3.14. Spiral Sink

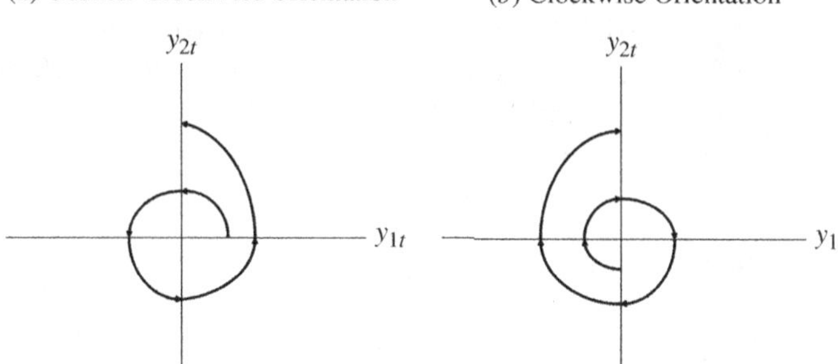

Fig. 3.15. Spiral Source

3.4 Repeated Pairs of Complex Eigenvalues

Consider the multi-dimensional, linear dynamical system

$$x_{t+1} = Ax_t + B. \tag{3.54}$$

Suppose that the n x n matrix of coefficients, A, has $n/2$ pairs of repeated complex eigenvalues $\{\mu, \bar{\mu}, \mu, \bar{\mu}, \cdots \mu, \bar{\mu}\}$, where

$$\mu \equiv \alpha + \beta i,$$
$$\bar{\mu} \equiv \alpha - \beta i, \tag{3.55}$$

and $i \equiv \sqrt{-1}$. Suppose further that $|I - A| \neq 0.$[2]

As established in Lemma 2.7 and (2.88), there exists a time-invariant transformation of the vector of original state variables, x_t, into a dynamical system of interdependent state variables, y_t, whose evolution can be characterized based on the properties of the Jordan matrix.

In particular, there exists a nonsingular n x n matrix, Q, such that

$$x_t = Qy_t + \bar{x}, \tag{3.56}$$

where $\bar{x} = [I - A]^{-1}B$ is the steady-state equilibrium of the system, and

$$y_{t+1} = Dy_t, \tag{3.57}$$

---

[2] If n were odd, then the additional eigenvalue would necessarily be real.

where

$$D = \begin{bmatrix} \alpha & -\beta & 0 & 0 & 0 & 0 & 0 & 0 \\ \beta & \alpha & 0 & 0 & 0 & 0 & 0 & 0 \\ 1 & 0 & \alpha & -\beta & 0 & 0 & 0 & 0 \\ 0 & 1 & \beta & \alpha & 0 & 0 & 0 & 0 \\ \vdots & \vdots & \ddots & \ddots & \ddots & \ddots & \vdots & \vdots \\ \vdots & \vdots & \ddots & \ddots & \ddots & \ddots & \vdots & \vdots \\ 0 & 0 & 0 & 0 & 1 & 0 & \alpha & -\beta \\ 0 & 0 & 0 & 0 & 0 & 1 & \beta & \alpha \end{bmatrix}. \tag{3.58}$$

Hence,

$$y_t = D^t y_0. \tag{3.59}$$

Thus, for $j = 1, 2, \cdots, n/2$,

$$
\begin{aligned}
y_{2j-1,t} &= \sum_{k=0}^{j-1} r^{t-k} \binom{t}{k} [\cos(t-k)\theta y_{2(j-k)-1,0} \\
&\quad - \sin(t-k)\theta y_{2(j-k),0}] \\
y_{2j,t} &= \sum_{k=0}^{j-1} r^{t-k} \binom{t}{k} [\sin(t-k)\theta y_{2(j-k)-1,0} \\
&\quad + \cos(t-k)\theta y_{2(j-k),0}].
\end{aligned}
\tag{3.60}
$$

Since $x_t = Qy_t + \bar{x}$, the evolution of each of the original state variables, x_{it}, $i = 1, 2, ..., n$, is given by

$$x_{it} = \sum_{m=0}^{(n/2)-1} r^{t-m} \binom{t}{m} [K_{im}\cos(t-m)\theta + \tilde{K}_{im}\sin(t-m)\theta] + \bar{x}_i, \tag{3.61}$$

where K_{im} and $\tilde{K}_{im}$ are constants.

Theorem 3.5. *(Necessary and Sufficient Conditions for Global Stability: Repeated Complex Eigenvalues)*
Consider the system $x_{t+1} = Ax_t + B$, *where* $x_t \in \Re^n$, *Suppose that* $|I - A| \neq 0$ *and suppose that* A *has* $n/2$ *pairs of repeated complex eigenvalues* $\{\mu, \overline{\mu}, \mu, \overline{\mu}, \cdots\}$, *where* $\mu \equiv \alpha + \beta i$, $\overline{\mu} \equiv \alpha - \beta i$, *and* $i \equiv \sqrt{-1}$. *Then the steady-state equilibrium* $\overline{x} = [I - A]^{-1}B$ *is globally stable if and only if*

$$r \equiv \sqrt{(\alpha^2 + \beta^2)} < 1.$$

Proof: Since for all $t - m$, $0 \leq |\cos(t - m)\theta| \leq 1$ and $0 \leq |\sin(t - m)\theta| \leq 1$, it follows from (3.52) that $\lim_{t \to \infty} x_{it} = \overline{x}_i$ if and only if $r < 1$. $\qquad\qquad\square$

3.5 The General Case

The analysis in Sects. 3.1–3.4 analyzes the trajectories of dynamical systems in which the matrix of coefficients A has: (a) distinct real eigenvalues, (b) repeated real eigenvalues, (c) distinct complex eigenvalues, and (d) repeated complex eigenvalues. This analysis could be generalized and conditions could be placed on the modulus of the eigenvalues so as to assure that the system is globally stable regardless of the type of eigenvalues that characterize the matrix of coefficients A.

Corollary 3.6. *Consider the system* $x_{t+1} = Ax_t + B$, *where* $x_t \in \Re^n$, *and suppose that* $|I - A| \neq 0$. *Then, the steady-state equilibrium* $\overline{x} = [I - A]^{-1}B$ *is globally (asymptotically) stable if and only if the modulus of each eigenvalue of the matrix* A *is smaller than* 1.

In the two-dimensional case, as depicted in Fig. 3.16, a steady-state equilibrium of a two-dimensional linear system is globally (asymptotically) stable if the eigenvalues of the matrix of coefficients A are within the interior of the *unit disk*.

In particular, if the eigenvalues are real, then they ought to be in the open interval $(-1, 1)$ along the real axis, whereas if they are complex, their modulus, $[\alpha^2 + \beta^2]^{1/2}$, ought to be smaller than 1, namely (α, β) is within the unit disk.

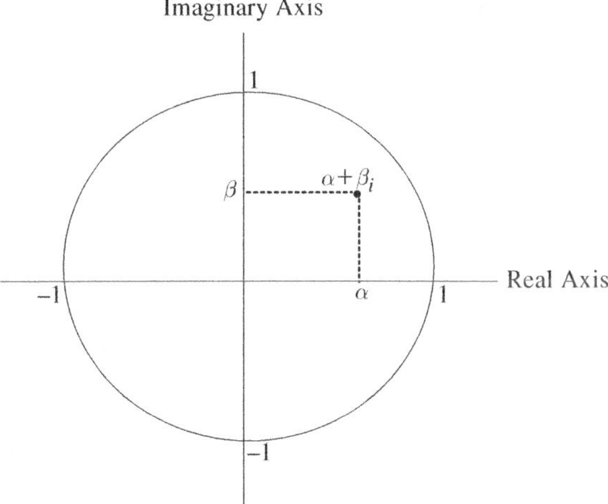

Fig. 3.16. Necessary and Sufficient Conditions for Global(Asymptotic) Stability of a Linear System: Eigenvalues within the Interior of the Unit Disk

3.6 Characterization of Two-Dimensional Systems in Terms of $tr\,A$ and det A

The qualitative properties of a two-dimensional dynamical system can be classified according to the relative values of the trace of the matrix of coefficients, $tr\ A$, and its determinant, $\det A$.

The eigenvalues of the matrix A are obtained as a solution to the equation

$$|A - \lambda I| = 0, \tag{3.62}$$

where $|A - \lambda I|$ is the determinant of the matrix $[A - \lambda I]$ and I is the identity matrix.

In the two-dimensional case, the eigenvalues, λ_1 and λ_2, are therefore obtained as a solution to the equation

$$\begin{vmatrix} a_{11} - \lambda & a_{21} \\ a_{12} & a_{22} - \lambda \end{vmatrix} = 0. \tag{3.63}$$

The implied characteristic polynomial is therefore

$$c(\lambda) \equiv \lambda^2 - tr A\lambda + \det A = 0, \tag{3.64}$$

and the eigenvalues, λ_1 and λ_2, are

$$\lambda_{1,2} = \frac{tr\,A \pm \sqrt{(tr\,A)^2 - 4\det A}}{2}. \tag{3.65}$$

Hence, the eigenvalues are real or complex depending on the relative value of $tr\ A$ and $\det A$.

$$\lambda_{1,2} \begin{cases} \text{complex} & if\ (tr A)^2 < 4\det A \\[2ex] \text{real} & if\ (tr A)^2 \geq 4\det A. \end{cases} \tag{3.66}$$

Proposition 3.7. *In a two-dimensional, first-order, linear system* $x_{t+1} = Ax_t + B$, *where* $x_t \in \Re^2$, *if the eigenvalues are real and distinct, i.e. if*

$$(tr A)^2 > 4\det A,$$

and thus $\lambda_1 > \lambda_2$, *then a steady-state equilibrium is a:*

- *Saddle (i.e.* $\{\lambda_1 > 1\ and\ |\lambda_2| < 1\}\ or\ \{|\lambda_1| < 1\ and\ \lambda_2 < -1\})$

  if and only if (as depicted in Fig. 3.17)

  $$\{c(1) < 0\ and\ c(-1) > 0\}$$

  or

  $$\{c(1) > 0\ and\ c(-1) < 0\},$$

  i.e.

  if and only if (as depicted in Fig. 3.19)

  $$-tr A - 1 < \det A < tr A - 1$$

  or

  $$\{tr A - 1 < \det A < -tr A - 1\}.$$

- *Stable Node (i.e.* $|\lambda_i| < 1$, $i = 1, 2$)

  if and only if (as depicted in Fig. 3.18)

  $$\{c(1) > 0\ and\ c(-1) > 0\},$$

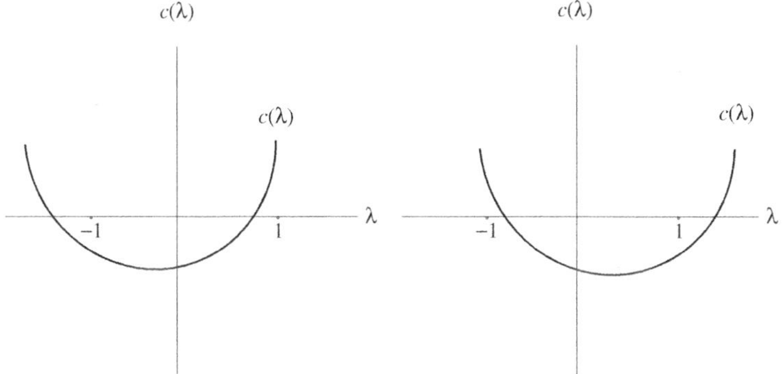

Fig. 3.17. The Characteristic Polynomial, $c(\lambda)$
A Saddle

i.e.

if and only if (as depicted in Fig. 3.19)

$$\{\det A > tr\,A - 1 \ \ and \ \ \det A > -tr\,A - 1\}.$$

- *Unstable Node (i.e. $|\lambda_i| > 1$, $i = 1, 2$)*

  if and only if (as depicted in Fig. 3.18)

$$\{c(1) < 0 \ \ and \ \ c(-1) < 0\},$$

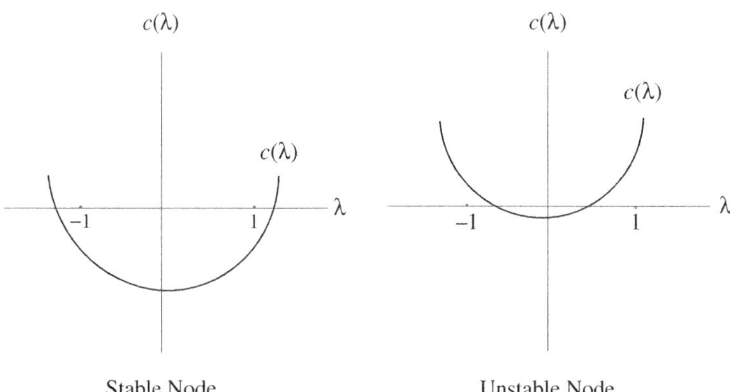

Fig. 3.18. The Characteristic Polynomial, $c(\lambda)$

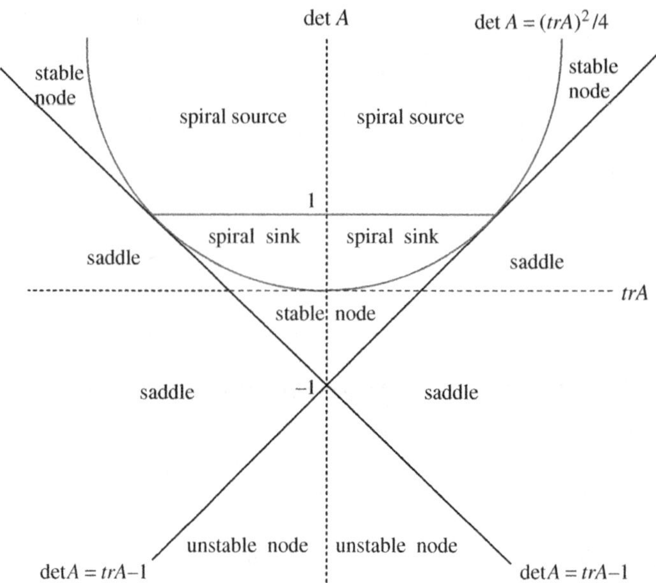

Fig. 3.19. Characterization of Steady-State Equilibria of Two-Dimensional Systems in Terms of the Trace, trA, and the Determinant, $\det A$, of the Matrix of Coefficients A

i.e.

if and only if (as depicted in Fig. 3.19)

$$\{\det A < trA - 1 \ and \ \det A < -trA - 1\}.$$

Proof: Follows from Figs. 3.17–3.19 and the evaluation of the characteristic polynomial $c(\lambda)$ at 1 and -1.   □

Proposition 3.8. *In a two-dimensional linear first-order system $x_{t+1} = Ax_t + B$, where $x_t \in \Re^2$, if the eigenvalues are complex, i.e. if*

$$(trA)^2 < 4\det A,$$

then the dynamical system is characterized by:

- *Spiral Sink*
  if and only if (as depicted in Fig. 3.19)

$$\det A < 1.$$

- *Spiral Source*
  if and only if (as depicted in Fig. 3.19)

$$\det A > 1.$$

Proof: Since the eigenvalues are complex and thus $(tr\ A)^2 < 4\det A$, it follows that

$$\mu_1 = \frac{tr\,A}{2} + \frac{\sqrt{4\det A - (tr\,A)^2}}{2}\,i \qquad (3.67)$$

$$\bar{\mu}_1 = \frac{tr\,A}{2} - \frac{\sqrt{4\det A - (tr\,A)^2}}{2}\,i.$$

Hence, the modulus of the eigenvalue, r, is

$$r = \left(\frac{tr\,A}{2}\right)^2 + \left(\frac{\sqrt{4\det A - (tr\,A)^2}}{2}\right)^2 = \det A. \qquad (3.68)$$

Thus, the steady-state equilibrium is globally stable and convergence is spiral if and only if $\det A < 1$, whereas the steady-state equilibrium is unstable and divergence is spiral if and only if $\det A > 1$.

$\square$

4

Multi-Dimensional, First-Order, Nonlinear Systems

This chapter characterizes the evolution of a vector of state variables in multi-dimensional, first-order, nonlinear systems of difference equations. It utilizes the analysis of linear, multi-dimensional, first-order systems to characterize the trajectory of nonlinear systems via their linearization in the proximity of a steady-state equilibrium, and the examination of the local and the global properties of these systems, based on the *Stable Manifold Theorem.*

The analysis examines the properties of the locally stable and unstable manifolds and the corresponding globally stable and unstable manifolds, and it analyzes the stability of the system based on the characterization of the linear case in Chaps. 2 and 3.

Consider the system of autonomous nonlinear first-order difference equations, where the evolution of the vector of state variables, x_t, is governed by the nonlinear system

$$x_{t+1} = \phi(x_t), \qquad t = 0, 1, 2, \cdots, \tag{4.1}$$

where $\phi : \Re^n \to \Re^n$. Namely,

$$
\begin{aligned}
x_{1t+1} &= \phi^1(x_{1t}, x_{2t}, \cdots, x_{nt}) \\
x_{2t+1} &= \phi^2(x_{1t}, x_{2t}, \cdots, x_{nt}) \\
&\;\;\vdots \qquad\qquad \vdots \\
x_{nt+1} &= \phi^n(x_{1t}, x_{2t}, \cdots, x_{nt}),
\end{aligned}
\tag{4.2}
$$

where $\phi^i : \Re^n \to \Re$, $i = 1, 2, ..., n$, is a continuously differentiable single-value function, and the initial value of the vector of state variables, $x_0 = (x_{10}, x_{20}, ..., x_{n0})$, is given.[1]

A solution to the difference equation $x_{t+1} = \phi(x_t)$ is a *trajectory* (or an *orbit*) of the vector of state variables, $\{x_t\}_{t=0}^{\infty}$, that satisfies this law of motion at any point in time. It relates the value of the vector of state variables at time t, x_t, to its initial value, x_0, based on the function $\phi(x_t)$.

Steady-state equilibria provide essential reference points for the characterization of nonlinear dynamical systems. A *steady-state equilibrium* (alternatively defined as a *stationary equilibrium*, a *rest point*, an *equilibrium point*, or a *fixed point*) is a value of the vector of state variables, x_t, that is invariant under the law of motion dictated by the dynamical system.

Definition 4.1. *(A Steady-State Equilibrium)*
A steady-state equilibrium of the nonlinear system of difference equations $x_{t+1} = \phi(x_t)$ *is a vector* $\bar{x} \in \Re^n$ *such that*

$$\bar{x} = \phi(\bar{x}).$$

Generically, a nonlinear system may be characterized by the existence of a unique steady-state equilibrium, the existence of multiplicity of (distinct) steady-state equilibria, the existence of chaotic behavior, or the non-existence of a steady-state equilibrium. Furthermore, the nonlinear system may converge to a steady-state equilibrium, may diverge to plus or minus infinity, may converge to a periodic orbit, and, unlike a linear system, a nonlinear system may exhibit a chaotic behavior. Hence, a qualitative examination of a dynamical system requires the analysis of the asymptotic behavior of the system as time approaches infinity.

The characterization of the qualitative behavior of this nonlinear dynamical system requires its linear approximation in the vicinity of its steady-state equilibrium, $\bar{x}$. In particular, the stability analysis of the system's steady-state equilibria determines whether a steady-state equilibrium is attractive or repulsive for all or at least some set of initial conditions. It facilitates the study of the local, and often the global, properties of a dynamical system, and it permits analysis of the implications of small, and sometimes large, perturbations that occur once the system is in the vicinity of a steady-state equilibrium.

---

[1] For local analysis, it is sufficient that the function $\phi^i : \Re^n \to \Re$ be continuously differentiable only in some neighborhood of the relevant steady-state equilibrium.

If for a sufficiently small perturbation the dynamical system converges asymptotically to the original equilibrium, the system is *locally* stable, whereas if regardless of the magnitude of the perturbation the system converges asymptotically to the original equilibrium, the system is *globally* stable. Formally, the definitions of local and global stability are as follows:[2]

Definition 4.2. *(Local and Global Stability of a Steady-State Equilibrium)*
A steady-state equilibrium, $\overline{x}$, of the nonlinear system $x_{t+1} = \phi(x_t)$ is:

- *globally (asymptotically) stable, if*

$$\lim_{t \to \infty} x_t = \overline{x} \quad \forall x_0 \in \Re^n$$

- *locally (asymptotically) stable, if*

$$\exists \epsilon > 0 \ such \ that \ \lim_{t \to \infty} x_t = \overline{x} \quad \forall x_0 \in B_\epsilon(\overline{x}),$$

where $B_\epsilon(\overline{x}) \equiv \{x \in \Re^n : |x_i - \bar{x}_i| < \varepsilon \ \forall i = 1, 2, 3, ...n\}.$

Thus, a steady-state equilibrium is *globally* (asymptotically) stable if the system converges to this steady-state equilibrium regardless of the initial condition, whereas a steady-state equilibrium is *locally* (asymptotically) stable if there exists an ϵ- neighborhood of the steady-state equilibrium such that from every initial condition within this neighborhood the system converges to this steady-state equilibrium.

Global stability of a steady-state equilibrium necessitates global uniqueness of the steady-state equilibrium. Clearly, if there is more than one steady-state equilibrium, none of the equilibria can be globally stable since if the systems is in one steady state equilibrium other steady-state equilibria will never be reached.

Local stability of a steady-state equilibrium necessitates local uniqueness of the steady-state equilibrium. Namely, it requires the absence of any additional point in a close neighborhood of the steady-state from which there is no escape. Clearly, if the system is characterized by a

---

[2] The economic literature, to a large extent, refers to the stability concepts in Definition 4.2 as global stability and local stability, respectively, whereas the mathematical literature refers to them as global asymptotic stability and local asymptotic stability, respectively. The concept of stability in the mathematical literature is reserved to situations in which trajectories that are initiated from an $\epsilon-$neighborhood of a fixed point remain sufficiently close to this fixed point thereafter.

continuum of equilibria, none of these steady-state equilibria is locally stable. There exists no neighborhood of a steady-state equilibrium that does not contain additional steady-state equilibria, and hence there exist initial conditions within an ε- neighborhood of a steady-state equilibrium that do not lead to this steady-state equilibrium in the long run. Thus, local stability of a steady-state equilibrium requires that this equilibrium be locally unique.

4.1 Local Analysis

The dynamical system is characterized initially in the proximity of a steady-state equilibrium.

4.1.1 Linearization

Suppose that the dynamical system has a steady-state equilibrium, $\bar{x}$. Namely $\exists \bar{x} \in \Re^n$ such that $\bar{x} = \phi(\bar{x})$.

The function $x_{t+1} = \phi(x_t)$ can be approximated around the steady-state value, $\bar{x}$. In particular, a Taylor expansion of $x_{it+1} = \phi^i(x_t)$, $i = 1, 2, ..., n$, around the steady-state value, $\bar{x}$, yields

$$x_{it+1} = \phi^i(x_t) = \phi^i(\bar{x}) + \sum_{j=1}^{n} \frac{\partial \phi^i(\bar{x})}{\partial x_{jt}}(x_{jt} - \bar{x}_j) + \cdots + R_n, \qquad (4.3)$$

where R_n is a residual term.

Thus, the linearized equation around the steady-state, $\bar{x}$, is defined by[3]

$$x_{it+1} = \frac{\partial \phi^i(\bar{x})}{\partial x_{1t}} x_{1t} + \frac{\partial \phi^i(\bar{x})}{\partial x_{2t}} x_{2t} + \cdots + \frac{\partial \phi^i(\bar{x})}{\partial x_{nt}} x_{nt}$$

$$+ \phi^i(\bar{x}) - \sum_{j=1}^{n} \frac{\partial \phi^i(\bar{x})}{\partial x_{jt}} \bar{x}_j. \qquad (4.4)$$

The linearized system is therefore

---

[3] Since higher order terms (i.e., $(x_{jt} - \bar{x}_j)^k$ for $k \geq 2$) are ignored, this approximation becomes increasingly less accurate the further the systems is from the steady-state equilibrium, $\bar{x}$.

$$
\begin{bmatrix} x_{1t+1} \\ x_{2t+1} \\ \vdots \\ x_{nt+1} \end{bmatrix} = \begin{bmatrix} \dfrac{\partial \phi^1(\overline{x})}{\partial x_{1t}} & \dfrac{\partial \phi^1(\overline{x})}{\partial x_{2t}} & \cdots & \dfrac{\partial \phi^1(\overline{x})}{\partial x_{nt}} \\ \dfrac{\partial \phi^2(\overline{x})}{\partial x_{1t}} & \dfrac{\partial \phi^2(\overline{x})}{\partial x_{2t}} & \cdots & \dfrac{\partial \phi^2(\overline{x})}{\partial x_{nt}} \\ \vdots & \vdots & & \vdots \\ \dfrac{\partial \phi^n(\overline{x})}{\partial x_{1t}} & \dfrac{\partial \phi^n(\overline{x})}{\partial x_{2t}} & \cdots & \dfrac{\partial \phi^n(\overline{x})}{\partial x_{nt}} \end{bmatrix} \begin{bmatrix} x_{1t} \\ x_{2t} \\ \vdots \\ x_{nt} \end{bmatrix}
\tag{4.5}
$$

$$
+ \begin{bmatrix} \phi^1(\overline{x}) - \sum_{j=1}^{n} \dfrac{\partial \phi^1(\bar{x})}{\partial x_{jt}} \overline{x}_j \\ \phi^2(\overline{x}) - \sum_{j=1}^{n} \dfrac{\partial \phi^2(\bar{x})}{\partial x_{jt}} \overline{x}_j \\ \vdots \\ \phi^n(\overline{x}) - \sum_{j=1}^{n} \dfrac{\partial \phi^n(\bar{x})}{\partial x_{jt}} \overline{x}_j \end{bmatrix} .
$$

Thus, the nonlinear system is approximated locally (around a steady-state equilibrium) by a linear system,

$$
x_{t+1} = A x_t + B,
\tag{4.6}
$$

where

$$
A \equiv \begin{bmatrix} \dfrac{\partial \phi^1(\overline{x})}{\partial x_{1t}} & \dfrac{\partial \phi^1(\overline{x})}{\partial x_{2t}} & \cdots & \dfrac{\partial \phi^1(\overline{x})}{\partial x_{nt}} \\ \dfrac{\partial \phi^2(\overline{x})}{\partial x_{1t}} & \dfrac{\partial \phi^2(\overline{x})}{\partial x_{2t}} & \cdots & \dfrac{\partial \phi^2(\overline{x})}{\partial x_{nt}} \\ \vdots & \vdots & & \vdots \\ \dfrac{\partial \phi^n(\overline{x})}{\partial x_{1t}} & \dfrac{\partial \phi^n(\overline{x})}{\partial x_{2t}} & \cdots & \dfrac{\partial \phi^n(\overline{x})}{\partial x_{nt}} \end{bmatrix} \equiv \mathcal{D}\phi(\overline{x}),
\tag{4.7}
$$

is the Jacobian matrix of $\phi(x_t)$ evaluated at $\overline{x}$, and

$$
B \equiv \begin{bmatrix} \phi^1(\overline{x}) - \sum_{j=1}^{n} \dfrac{\partial \phi^1(\bar{x})}{\partial x_{jt}} \overline{x}_j \\ \phi^2(\overline{x}) - \sum_{j=1}^{n} \dfrac{\partial \phi^2(\bar{x})}{\partial x_{jt}} \overline{x}_j \\ \vdots \\ \phi^n(\overline{x}) - \sum_{j=1}^{n} \dfrac{\partial \phi^n(\bar{x})}{\partial x_{jt}} \overline{x}_j \end{bmatrix} ,
\tag{4.8}
$$

is a constant column vector.

As established in the theorem below, the local behavior of the non-linear dynamical system in the proximity of a steady-state equilibrium, $\bar{x}$, can be assessed on the basis of the behavior of the linear system that approximates the nonlinear one in the vicinity of this steady-state equilibrium. Hence, the eigenvalues of the Jacobian matrix $\mathcal{D}\phi(\bar{x})$ determine the local behavior of the nonlinear system according to the results stated in Theorems 3.1–3.2 and Corollary 3.6.

4.1.2 Stable, Unstable, and Center Eigenspaces

The stable and unstable eigenspaces provide an essential reference point to the local characterization of a nonlinear dynamical system in the proximity of a steady-state equilibrium.

Definition 4.3. *(Stable, Unstable, and Center Eigenspaces)*
Let $\phi(x):\ \Re^n \to \Re^n$ be a continuously differentiable single-value function, and let $\mathcal{D}\phi(\bar{x})$ be the Jacobian matrix of $\phi(x)$ evaluated at a steady-state equilibrium, $\bar{x}$, i.e.

$$\mathcal{D}\phi(\bar{x}) = \begin{bmatrix} \dfrac{\partial \phi^1(\bar{x})}{\partial x_{1t}} & \dfrac{\partial \phi^1(\bar{x})}{\partial x_{2t}} & \cdots & \dfrac{\partial \phi^1(\bar{x})}{\partial x_{nt}} \\[2ex] \dfrac{\partial \phi^2(\bar{x})}{\partial x_{1t}} & \dfrac{\partial \phi^2(\bar{x})}{\partial x_{2t}} & \cdots & \dfrac{\partial \phi^2(\bar{x})}{\partial x_{nt}} \\[2ex] \vdots & \vdots & & \vdots \\[2ex] \dfrac{\partial \phi^n(\bar{x})}{\partial x_{1t}} & \dfrac{\partial \phi^n(\bar{x})}{\partial x_{2t}} & & \dfrac{\partial \phi^n(\bar{x})}{\partial x_{nt}} \end{bmatrix}.$$

- *The stable eigenspace, $E^s(\bar{x})$, of the steady-state equilibrium, $\bar{x}$, is*

$$E^s(\bar{x}) = span\{eigenvectors\ of\ \mathcal{D}\phi(\bar{x})\ whose\ eigenvalues$$
$$have\ modulus < 1\}.$$

- *The unstable eigenspace, $E^u(\bar{x})$, of the steady-state equilibrium, $\bar{x}$, is*

$$E^u(\bar{x}) = span\{eigenvectors\ of\ \mathcal{D}\phi(\bar{x})\ whose\ eigenvalues$$
$$have\ modulus > 1\}.$$

- *The center eigenspace, $E^c(\overline{x})$, of the steady-state equilibrium, $\overline{x}$, is*

$$E^c(\overline{x}) = span\{eigenvectors\ of\ \mathcal{D}\phi(\overline{x})\ whose\ eigenvalues$$

$$have\ modulus = 1\}.$$

Since the n eigenvectors of the $n \times n$ Jacobian matrix $\mathcal{D}\phi(\overline{x})$ span $\Re^n$, then as stated in the following corollary, the sum of the dimensions of the stable, unstable, and center eigenspaces is equal to the dimensionality of $\Re^n$.

Corollary 4.4. *Let $\mathcal{D}\phi(\overline{x})$ be the Jacobian matrix of $\phi(x)$ evaluated at a steady-state equilibrium $\overline{x}$.*

$$\dim E^s(\overline{x}) + \dim E^u(\overline{x}) + \dim E^c(\overline{x}) = n.$$

The *stable eigenspace* relative to the steady-state equilibrium, $\overline{x}$, is defined as the space spanned by the eigenvectors of $\mathcal{D}\phi(\overline{x})$ associated with eigenvalues of modulus smaller than one. Namely, the stable eigenspace is the geometric locus of all vectors, x_t, that upon a sufficient number of forward iterations under the map ϕ are mapped in the limit towards the steady-state equilibrium, $\overline{x}$.

The *unstable eigenspace* relative to the steady-state equilibrium, $\overline{x}$, is defined as the space spanned by the eigenvectors of $\mathcal{D}\phi(\overline{x})$ associated with eigenvalues of modulus larger than one. That is, the unstable eigenspace is the geometric locus of all vectors, x_t, that upon a sufficient number of backward iterations under the map ϕ are mapped in the limit to the steady-state equilibrium, $\bar{x}$.

The *center eigenspace* is the space spanned by the eigenvectors associated with eigenvalues of modulus equal to one. Namely, the center eigenspace is the geometric locus of all vectors, x_t, that are invariant under forward or backward iterations of the map ϕ.

As depicted in Fig. 4.1, if the nonlinear dynamical system is two-dimensional and the steady-state equilibrium is a saddle, then the stable eigenspace is one-dimensional and the unstable eigenspace is one-dimensional.

If the steady-state equilibrium is a stable node, then the stable eigenspace is two-dimensional and it consists of the entire two-dimensional real plane (i.e. $E^s(\overline{x}) = \Re^2$) and the dimensionality of the unstable eigenspace is zero. In contrast, if the steady-state equilibrium is an unstable node, then the unstable eigenspace is two-dimensional and it consists of the entire two-dimensional real plane (i.e. $E^u(\overline{x}) = \Re^2$). The dimensionality of the stable eigenspace is therefore zero.

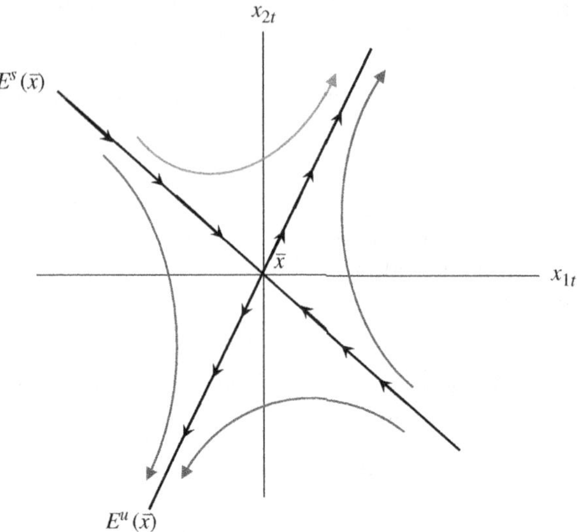

Fig. 4.1. Stable and Unstable Eigenspace
$\bar{x}$ is a Saddle Point

If the nonlinear dynamical system is two-dimensional and one of
the eigenvalues is equal to one, then the system is characterized by
a one-dimensional continuum of steady-state equilibria. If the second
eigenvalue is of modulus less than one, then, as depicted in Fig. 4.2 the

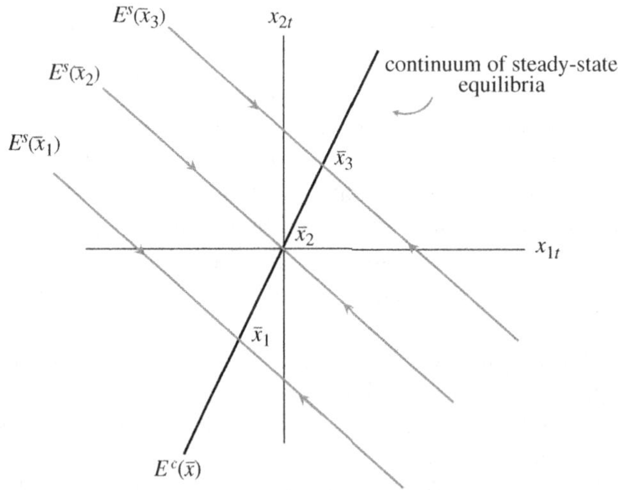

Fig. 4.2. A Center and a Stable Eigenspace
Continuum of Equilibria, $\bar{x}$

following holds: the stable eigenspace with respect to a given steady-state (e.g. $\bar{x}_1$) is one-dimensional, the center eigenspace is the one-dimensional continuum of steady-state equilibria, and the dimensionality of the unstable eigenspace is zero. If the second eigenvalue is of modulus greater than one, then the unstable eigenspace and the center eigenspace are each one-dimensional and the dimensionality of the stable eigenspace is zero.

In the subsequent analysis, it is assumed that all eigenvalues are of modulus different than one, and thus the center eigenspace is empty.[4]

Definition 4.5. *(Hyperbolic Fixed Point)*
Consider the map $\phi : \Re^n \to \Re^n$ *and let* $D\phi(\bar{x})$ *be the Jacobian matrix of* $\phi(x)$, *evaluated at a steady-state equilibrium* $\bar{x}$. *The steady-state equilibrium,* $\bar{x}$, *is a hyperbolic fixed point if* $D\phi(\bar{x})$ *has no eigenvalues of modulus one.*

4.1.3 Local Stable and Unstable Manifolds

The stable and unstable manifolds provide the nonlinear counterparts for the stable and unstable eigenspaces.[5]

Definition 4.6. *(Local Stable and Unstable Manifolds)*
Consider the nonlinear dynamical system

$$x_{t+1} = \phi(x_t).$$

- *A local stable manifold,* $W^s_{loc}(\bar{x})$, *of a steady-state equilibrium,* $\bar{x}$, *is*

$$W^s_{loc}(\bar{x}) = \{x \in U \,|\, \lim_{n \to +\infty} \phi^{\{n\}}(x)$$
$$= \bar{x} \text{ and } \phi^{\{n\}}(x) \in U \quad \forall n \in N\}.$$

- *A local unstable manifold,* $W^u_{loc}(\bar{x})$, *of a steady-state equilibrium,* $\bar{x}$, *is*

$$W^u_{loc}(\bar{x}) = \{x \in U \,|\, \lim_{n \to +\infty} \phi^{-\{n\}}(x)$$
$$= \bar{x} \text{ and } \phi^{\{n\}}(\bar{x}) \in U \quad \forall n \in N\},$$

---

[4] For a comprehensive exploration of the properties of the center manifold and its implications in the continuous time case, see Guckenheimer and Holms (1990).

[5] An n-dimensional manifold $M \subset \Re^N$ is a set such that $\forall x \in M$ there exists a neighborhood U for which there is a smooth invertible mapping (deffeomorphisim) $\phi : \Re^n \to U$ (for $n \leq N$).

where $U \equiv B_\epsilon(\overline{x})$ for some $\epsilon > 0$, and $\phi^{\{n\}}(x)$ is the nth iteration over x under the map ϕ.

Thus, a *local stable manifold* is the geometric place of all vectors $x \in \Re^n$ in some ball of radius ϵ, around the steady-state equilibrium, $\bar{x}$, whose elements asymptotically approach the steady-state equilibrium, $\overline{x}$, as the number of iterations under the map ϕ approaches infinity. Similarly, a *local unstable manifold* is the geometric place of all vectors $x \in \Re^n$ in some ball of radius ϵ around the steady-state equilibrium, $\bar{x}$, whose elements approach asymptotically the steady-state equilibrium, $\overline{x}$, as the number of backward iterations under the map ϕ approaches infinity.

4.1.4 The *Stable Manifold Theorem*

The *Stable Manifold Theorem* establishes the relationship between the stable and unstable eigenspaces and local stable and unstable manifolds in the proximity of a steady-state equilibrium.

Theorem 4.7. *(The Stable Manifold Theorem)*
Let $\phi : \Re^n \to \Re^n$ be a C^1 diffeomorphism[6] with a hyperbolic fixed point $\overline{x}$. Then there exist locally stable and unstable manifolds, $W^s_{loc}(\overline{x})$ and $W^u_{loc}(\overline{x})$, that are tangent, respectively, to the eigenspaces $E^s(\overline{x})$ and $E^u(\overline{x})$ of the Jacobian matrix, $D\phi(\overline{x})$, at $\bar{x}$. Furthermore,

$$\dim W^s_{loc}(\overline{x}) = \dim E^s(\overline{x})$$

$$\dim W^u_{loc}(\overline{x}) = \dim E^u(\overline{x}).$$

Proof. See Nitecki (1971).

Figure 4.3 depicts local stable and unstable manifolds, $W^s_{loc}(\overline{x})$ and $W^u_{loc}(\overline{x})$, and the stable and unstable eigenspaces, $E^s(\overline{x})$ and $E^u(\overline{x})$, in the vicinity of a steady-state equilibrium, $\bar{x}$, for the two-dimensional case in which the steady-state equilibrium is a saddle point. As established in the *Stable Manifold Theorem*, $W^s_{loc}(\overline{x})$ is tangent to $E^s(\overline{x})$ at the steady-state equilibrium, $\bar{x}$, and both are one-dimensional. Similarly, $W^u_{loc}(\overline{x})$ is tangent to $E^u(\overline{x})$ at the steady-state equilibrium, $\overline{x}$, and both are one-dimensional.

---

[6] A one to one mapping ϕ from $\Re^n$ onto itself, is a C^1 diffeomorphism if ϕ and ϕ^{-1} are continuously differentiable (Munkres (1999)).

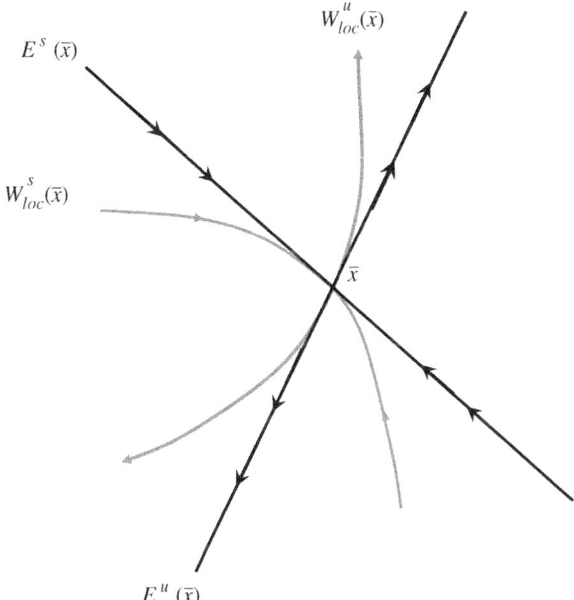

Fig. 4.3. The Local Stable and Unstable Manifolds, $W_{loc}^{s}(\bar{x})$ and $W_{loc}^{u}(\bar{x})$, in relation to the Stable and Unstable Eigenspaces, $E^{s}(\bar{x})$ and $E^{u}(\bar{x})$ $\bar{x}$ is a Saddle Point.

Hence Theorem 4.8 follows from Corollary 3.6 and the *Stable Manifold Theorem*.

Theorem 4.8. *(Sufficient Conditions for Local Stability of a Nonlinear System)*
Let $\phi : \Re^{n} \to \Re^{n}$ *be a* C^{1} *diffeomorphism with a hyperbolic fixed point,* $\bar{x}$. *Then a steady-state equilibrium,* $\bar{x}$, *is locally (asymptotically) stable if and only if the moduli of all eigenvalues of the Jacobian matrix,* $D\phi(\bar{x})$, *are smaller than* 1.

Remark. The evolution of the nonlinear system in the proximity of a steady-state equilibrium, $\bar{x}$, cannot be examined based on the linearized system, $D\phi(\bar{x})$, if the steady-state equilibrium is non-hyperbolic (i.e. if the modulus of one of the eigenvalues is equal to 1). The examination of this system would be based on the properties of the *center manifold*.[7]

---

[7] See Gukenhiemer and Holmes (1990) and Hale and Kocak (1991).

4.2 Global Analysis

Global analysis of a multi-dimensional nonlinear system can be advanced using the concepts of the *global stable manifold* and the *global unstable manifold.*

Definition 4.9. *(Globally Stable and Unstable Manifolds)*
Consider the nonlinear dynamical system

$$x_{t+1} = \phi(x_t),$$

and let $\overline{x}$ be the steady-state equilibrium of the system.

- *The global stable manifold, $W^s(\overline{x})$, of a steady-state equilibrium, $\overline{x}$, is*

$$W^s(\overline{x}) = \cup_{n \in N}\{\phi^{-\{n\}}(W^s_{loc}(\overline{x}))\}.$$

- *The global unstable manifold, $W^u(\overline{x})$, of a steady-state equilibrium, $\overline{x}$, is*

$$W^u(\overline{x}) = \cup_{n \in N}\{\phi^{\{n\}}(W^u_{loc}(\overline{x}))\}.$$

Thus, the *global stable manifold* is obtained by the union of all backward iteration under the map ϕ over the local stable manifold, and the global unstable manifold is obtained by the union of all forward iterations under the map ϕ over the local unstable manifold.

In the context of a two-dimensional dynamical system the tracing of the global stable and unstable manifolds based upon the properties of the local stable and unstable manifolds permits a global characterization of the dynamical system based on the local behavior of the system. In particular, the global properties of the dynamical system can be inferred from the properties of the local stable and unstable manifolds.

Figure 4.4 depicts the global stable and unstable manifolds, $W^s(\overline{x})$ and $W^u(\overline{x})$, in relation to the local stable and unstable manifolds, $W^s_{loc}(\overline{x})$ and $W^u_{loc}(\overline{x})$, and the stable and unstable eigenspaces, $E^s(\overline{x})$ and $E^u(\overline{x})$, in the vicinity of the steady-state equilibrium, $\bar{x}$, for the two-dimensional case in which the steady-state equilibrium is a saddle point.

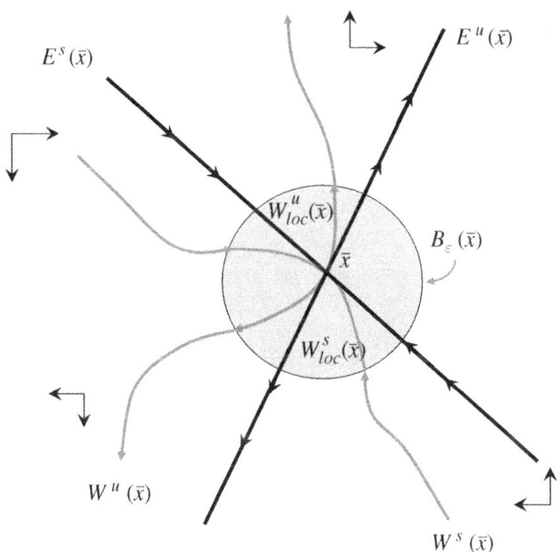

Fig. 4.4. The Global Stable and Unstable Manifolds, $W^s(\bar{x})$ and $W^u(\bar{x})$, in relation to the Local Stable and Unstable Manifolds, $W^s_{loc}(\bar{x})$ and $W^u_{loc}(\bar{x})$, and the Stable and Unstable Eigenspaces, $E^s(\bar{x})$ and $E^u(\bar{x})$
$\bar{x}$ is a Saddle Point

Theorem 4.11 provides a very restrictive sufficient condition for global stability that is unlikely to be satisfied by a conventional economic system. In light of the *Contraction Mapping Theorem*, the sufficient conditions for global stability in the one-dimensional case (Corollary 1.14) can be generalized for multi-dimensional dynamical systems. Analogously to the one-dimensional case, a contraction mapping in the n-dimensional case is defined as follows:

Definition 4.10. *(Contraction Mapping)*
Let $(\Re^n, \rho)$ *be a metric space. Then* $\phi(x) : \Re^n \to \Re^n$ *is a contraction mapping if for some* $\beta \in (0,1)$,

$$\rho(\phi(x^1), \phi(x^2)) \le \beta\rho(x^1, x^2) \qquad \forall x^1, x^2 \in \Re^n,$$

where $\rho(c,d) \equiv |c - d|$.

Theorem 4.11. *(Sufficient Conditions for Global Stability of a Nonlinear System)*
A stationary equilibrium of the multi-dimensional, autonomous, first-order difference equation, $x_{t+1} = \phi(x_t)$ *exists, is unique, and is globally stable if* $\phi : \Re^n \to \Re^n$ *is a contraction mapping.*

5

Higher-Order and Non-Autonomous Systems

5.1 Higher-Order Systems

This chapter characterizes the evolution of a vector of state variables in higher-order systems as well as non-autonomous systems. It establishes the solution method for these higher-order and non-autonomous systems and, it analyzes the factors that determine the qualitative properties of these discrete dynamical systems in the linear and subsequently the nonlinear case.

The analysis is based upon the transformation of higher-order systems and non-autonomous systems into a multi-dimensional first-order system that can be examined based on the analysis in Chaps. 2–4. In particular, a one-dimensional second-order system is converted into a two-dimensional first-order system, a one-dimensional third-order system is transformed into a three-dimensional first-order system, a one-dimensional n^{th}-order system is converted into an n-dimensional first-order system, and an n-dimensional m^{th}-order system is transformed into an n x m-dimensional first-order system. Similarly, the analysis of non-autonomous systems is based on their transformation into higher-order, time-independent (autonomous) systems that can be examined based on the analysis of multi-dimensional first-order systems in Chaps. 2–4.

5.1.1 Linear Systems

Second-Order Systems

Consider a one-dimensional, *second-order*, autonomous, linear difference equation that governs the evolution of a one-dimensional state variable, x_t, over time. Unlike the first-order case, the value of the state variable at time $t+2$ depends upon its value at time $t+1$ and at time t. In particular, consider the difference equation

$$x_{t+2} + a_1 x_{t+1} + a_0 x_t + b = 0, \tag{5.1}$$

where the value of the state variable at time t, x_t, is a real number, i.e., $x_t \in \Re$, the constant parameters a_0, a_1, and b are real numbers, i.e., $a_0, a_1, b \in \Re$, and the initial values of the state variable at time 0, x_0, and time $1, x_1$, are given.

In order to examine this difference equation in a familiar manner, using the basic propositions established in Chaps. 2–4, this one-dimensional second-order difference equation is converted into a two-dimensional first-order system.

Define a new state variable, y_t, such that,

$$x_{t+1} \equiv y_t, \tag{5.2}$$

and therefore

$$x_{t+2} = y_{t+1}. \tag{5.3}$$

Substituting (5.2) and (5.3) into (5.1), the one-dimensional, second-order difference equation can be transformed into the two-dimensional first-order system,

$$\begin{cases} y_{t+1} + a_1 y_t + a_0 x_t + b = 0 \\ x_{t+1} = y_t, \end{cases} \tag{5.4}$$

or

$$\begin{cases} y_{t+1} = -a_1 y_t - a_0 x_t - b \\ x_{t+1} = y_t. \end{cases} \tag{5.5}$$

Hence, the one-dimensional, second-order difference equation is converted into a two-dimensional first order system that characterizes the evolution of two state variables, $\{y_t, x_t\}$, over time.

$$\begin{bmatrix} y_{t+1} \\ x_{t+1} \end{bmatrix} = \begin{bmatrix} -a_1 & -a_0 \\ 1 & 0 \end{bmatrix} \begin{bmatrix} y_t \\ x_t \end{bmatrix} + \begin{bmatrix} -b \\ 0 \end{bmatrix}, \tag{5.6}$$

where the initial conditions of the two state variables, (y_0, x_0), are given since (x_0, x_1) are given, and as follow from (5.2), $y_0 = x_1$.

Thus, the characterization of a one-dimensional, second-order linear difference equation can be obtained based on the methods developed in Chaps. 2 and 3, and in particular the results stated in Theorems

3.1–3.2 and Corollary 3.6, once the difference equation is transformed into a two-dimensional, first-order, linear system, as analyzed above.

Third-Order Systems

Consider a one-dimensional, *third-order*, autonomous, linear difference equation that governs the evolution of a one-dimensional state variable, x_t, over time. The value of the state variable at time $t + 3$ depends on its value at time $t + 2$, $t + 1$, and t. In particular, consider the difference equation

$$x_{t+3} + a_2 x_{t+2} + a_1 x_{t+1} + a_0 x_t + b = 0, \tag{5.7}$$

where the value of the *state variable* at time t, x_t, is a real number, i.e. $x_t \in \Re$, the constant parameters $a_0, a_1, a_2, b \in \Re$, and the initial conditions of the system, (x_0, x_1, x_2), are given.

Define a new state variable, y_t, such that

$$x_{t+1} \equiv y_t, \tag{5.8}$$

and similarly, define an additional state variable z_t, such that

$$x_{t+2} = y_{t+1} \equiv z_t. \tag{5.9}$$

Then the third-order difference equation can be transformed into a three-dimensional, first-order system. In particular, since $x_{t+2} = y_{t+1} = z_t$, it follows that

$$x_{t+3} = y_{t+2} = z_{t+1}. \tag{5.10}$$

Substituting (5.8)–(5.10) into (5.7), the one-dimensional, third-order difference equation can be transformed into the three-dimensional first-order linear system,

$$\begin{cases} z_{t+1} + a_2 z_t + a_1 y_t + a_0 x_t + b = 0 \\ y_{t+1} = z_t \\ x_{t+1} = y_t, \end{cases} \tag{5.11}$$

or

$$\begin{cases} z_{t+1} = -a_2 z_t - a_1 y_t - a_0 x_t - b \\ y_{t+1} = z_t \\ x_{t+1} = y_t. \end{cases} \tag{5.12}$$

Thus, the one-dimensional, third-order difference equation is converted into a three-dimensional, first-order system that describes the evolution of three state variables, $\{y_t, z_t, x_t\}$, over time:

$$\begin{bmatrix} z_{t+1} \\ y_{t+1} \\ x_{t+1} \end{bmatrix} = \begin{bmatrix} -a_2 & -a_1 & -a_0 \\ 1 & 0 & 0 \\ 0 & 1 & 0 \end{bmatrix} \begin{bmatrix} z_t \\ y_t \\ x_t \end{bmatrix} + \begin{bmatrix} -b \\ 0 \\ 0 \end{bmatrix}. \tag{5.13}$$

where the initial conditions of the three state variables, (z_0, y_0, x_0), are given, noting that as follows from (5.8) and (5.9), $y_0 = x_1$ and $z_0 = x_2$.

Thus, the characterization of a one-dimensional, third-order linear system can be obtained based on the methods developed in Chaps. 2 and 3 and in particular the results stated in Theorems 3.1–3.2 and Corollary 3.6, once the system is transformed into a three-dimensional, first-order, linear system, as analyzed above.

$\mathbf{N}^{th}$-Order System

Consider a one-dimensional, $n^{th}\text{-}order$, autonomous, linear difference equation that governs the evolution of a one-dimensional state variable, x_t, over time. The value of the state variable at time $t+n$ depends upon its value at time $t+n-1$, $t+n-2$, ..., $t+2$, $t+1$ and t. In particular, consider the difference equation

$$x_{t+n} + a_{n-1}x_{t+n-1} + \cdots + a_1 x_{t+1} + a_0 x_t + b = 0, \tag{5.14}$$

where the value of the *state variable* at time t, $x_t \in \Re$, the constant parameters $a_0, a_1, a_2, \ldots a_{n-1}, b \in \Re$, and the initial conditions $(x_0, x_1, x_2, \ldots, x_{n-1})$ are given.

Define $n-1$ new state variables, such that

$$\begin{aligned} x_{t+1} &\equiv y_{1,t} \\ x_{t+2} &= y_{1,t+1} \equiv y_{2,t} \\ x_{t+3} &= y_{1,t+2} = y_{2,t+1} \equiv y_{3,t} \\ &\vdots \qquad \vdots \\ x_{t+n-1} &= y_{1,t+n-2} = y_{2,t+n-3} = \cdots = y_{n-2,t+1} \equiv y_{n-1,t}. \end{aligned} \tag{5.15}$$

It follows that

$$x_{t+n} = y_{n-1,t+1}.$$ (5.16)

Substituting (5.15) and (5.16) into (5.14), the one-dimensional, n^{th}-order difference equation can be transformed into the n-dimensional, first-order system

$$
\begin{bmatrix}
y_{n-1,t+1} \\
y_{n-2,t+1} \\
y_{n-3,t+1} \\
\vdots \\
y_{1,t+1} \\
x_{t+1}
\end{bmatrix}
=
\begin{bmatrix}
-a_{n-1} & -a_{n-2} & \cdots & -a_2 & -a_1 & -a_0 \\
1 & 0 & 0 & 0 & 0 & 0 \\
0 & 1 & 0 & 0 & 0 & 0 \\
\vdots & & \ddots & \ddots & \ddots & \vdots \\
0 & 0 & 0 & 1 & 0 & 0 \\
0 & 0 & 0 & 0 & 1 & 0
\end{bmatrix}
\begin{bmatrix}
y_{n-1,t} \\
y_{n-2,t} \\
y_{n-3,t} \\
\vdots \\
y_{1,t} \\
x_t
\end{bmatrix}
+
\begin{bmatrix}
-b \\
0 \\
0 \\
\vdots \\
0 \\
0
\end{bmatrix},
$$ (5.17)

where the initial conditions of the n state variables, x_0, $y_{1,0} = x_1$, $y_{2,0} = x_2$, ..., $= y_{n-1,t} = x_{n-1}$, are given, noting (5.15).

Thus, the characterization of a one-dimensional, n^{th}-order linear system can be obtained based on the methods developed in Chaps. 2 and 3, and in particular the results stated in Theorems 3.1–3.2 and Corollary 3.6, once the difference equation is transformed into an n-dimensional, first-order, linear system, as analyzed above.

More generally an n-dimensional m^{th}-order system can be converted into a m x n dimensional first-order system and analyzed according to the methods developed in Chaps. 2 and 3.

5.1.2 Nonlinear Systems

Consider a one-dimensional, n^{th}-order, autonomous, *nonlinear* system that governs the evolution of a one-dimensional state variable, x_t, over time. The value of the state variable at time $t+n$ depends on its value in time $t+n-1$, $t+n-2$, ..., $t+2$, $t+1$, and t. In particular, consider the system

$$x_{t+n} = \phi(x_{t+n-1}, x_{t+n-2}, x_{t+n-3}, \cdots, x_{t+1}, x_t),$$ (5.18)

where $\phi : \Re^n \to \Re$ is a single-valued function, the value of the *state variable* at any time t, x_t, is a real number, i.e. $x_t \in \Re$, and the initial conditions of the system, $(x_0, x_1, x_2, ..., x_{n-1})$, are given.

Let

$$
\begin{aligned}
x_{t+1} &\equiv y_{1,t} \\
x_{t+2} &= y_{1,t+1} \equiv y_{2,t} \\
x_{t+3} &= y_{1,t+2} = y_{2,t+1} \equiv y_{3,t} \\
&\ \ \vdots \qquad \vdots \\
x_{t+n-1} &= y_{1,t+n-2} = y_{2,t+n-3} = \cdots = y_{n-2,t+1} \equiv y_{n-1,t}.
\end{aligned}
$$

(5.19)

Substituting (5.19) into (5.18), the one-dimensional, n^{th}-order nonlinear difference equation can be transformed into the n-dimensional, first-order, nonlinear system

$$
\begin{aligned}
y_{n-1,t+1} &= \phi(y_{n-1,t}, y_{n-2,t}, y_{n-3,t}, \cdots, x_t) \\
y_{n-2,t+1} &= y_{n-1,t} \\
&\ \ \vdots \qquad \vdots \\
y_{1,t+1} &= y_{1,t},
\end{aligned}
$$

(5.20)

where the initial conditions of the n state variables, x_0, $y_{1,0} = x_1$, $y_{2,0} = x_2$, ... $y_{n-1,t} = x_{n-1}$, are given.

Thus, the nth-order nonlinear difference equation can be represented as an n - dimensional, first-order nonlinear system that can be analyzed according to the methods developed in Chap. 4.

5.2 Non-Autonomous Systems

The characterization of non-autonomous systems (i.e. systems in which the functional relationship between the state variables changes over time) requires the transformation of the time-dependant system into a higher order time-independent system.

Consider a system of non-autonomous, first-order, linear difference equations

$$
x_{t+1} = A(t)x_t + B(t),
$$

(5.21)

and a non-autonomous nonlinear system

$$
x_{t+1} = f(x_t, t),
$$

(5.22)

where the vector of state variables, $x_t \in \Re^n$, $A(t)$ is an $n \times n$ matrix of time dependent coefficients of real numbers, $a_{ij}(t) \in \Re$, $i, j = 1, 2, ..., n$, and $B(t)$ is an n - dimensional time dependent vector with elements

$b_i(t) \in \Re$, $i = 1, 2, ..., n$. The initial value of the vector of state variables, $x_0 = (x_{10}, x_{20}, x_{30}, ..., x_{n0})$, is given.

A non-autonomous system can be converted into an autonomous one. Let $y_t \equiv t$. Then $y_{t+1} = t + 1 = y_t + 1$. Thus the linear system is transformed into

$$x_{t+1} = A(y_t)x_t + B(y_t)$$
$$y_{t+1} = y_t + 1,$$

(5.23)

and the nonlinear system is converted to

$$x_{t+1} = f(x_t, y_t)$$
$$y_{t+1} = y_t + 1.$$

(5.24)

Namely, the non-autonomous system is converted into a higher-dimensional autonomous system.

However, these systems have no steady-state equilibrium. There exists no $\overline{y} \in \Re$ such that $\overline{y} = \overline{y} + 1$, and thus neither the linear system nor the nonlinear system has a steady-state equilibrium.

The qualitative analysis provided by Chaps. 2–4, which is based on the behavior of the system in the vicinity of a steady-state equilibrium, is therefore not applicable for this dynamical system.

The characterization of these systems necessitates a redefinition of the state variables, so as to assure the existence of steady-state equilibria and thus to assure that the analysis of the previous chapters could be fully utilized. This transformation, however, would depend on the particular form of the dynamical system.

6

Examples of Two-Dimensional Systems

This chapter provides a complete characterization of several representative examples of two-dimensional dynamical systems. These examples include a first-order linear system with real eigenvalues, a first-order linear system with complex eigenvalues that exhibits periodic orbit, a first-order linear system with complex eigenvalues that exhibits a spiral sink, a first-order nonlinear system characterized by oscillatory convergence, and a second-order one-dimensional system that is converted into a first-order, two-dimensional system characterized by a continuum of equilibria and oscillatory divergence.

6.1 First-Order Linear Systems

6.1.1 Real, Distinct, Positive Eigenvalues

Consider the two-dimensional system of first-order homogenous linear difference equations

$$x_{t+1} \equiv \begin{bmatrix} x_{1t+1} \\ x_{2t+1} \end{bmatrix} = \begin{bmatrix} 4 & 1 \\ 7 & 2.5 \end{bmatrix} \begin{bmatrix} x_{1t} \\ x_{2t} \end{bmatrix} \equiv Ax_t, \qquad (6.1)$$

where $x_0 \equiv [x_{10}, x_{20}]$ is given.

The characterization of the time path of this two-dimensional system of interdependent state variables is based on the construction of a time-independent transformation that converts the system into a new dynamical system of independent state variables whose evolution can be derived based on the analysis of the one-dimensional case. Alternatively, the qualitative aspects of the trajectory of this system can be examined based on the derivation of the system's phase diagram without an explicit solution.

A. Derivation of an Explicit Solution

The Eigenvalues λ_1 and λ_2 of the Matrix A

The eigenvalues of the matrix A are obtained as a solution to the equation

$$|A - \lambda I| = 0, \tag{6.2}$$

where $|A - \lambda I|$ is the determinant of the matrix $[A - \lambda I]$ and I is the identity matrix. In the two-dimensional case, the eigenvalues, λ_1 and λ_2, are therefore obtained as a solution to the equation

$$\begin{vmatrix} a_{11} - \lambda & a_{21} \\ a_{12} & a_{22} - \lambda \end{vmatrix} = 0. \tag{6.3}$$

The implied characteristic polynomial is

$$c(\lambda) \equiv \lambda^2 - tr A \lambda + \det A = 0, \tag{6.4}$$

and the eigenvalues are therefore determined by the solution to

$$\begin{cases} \lambda_1 + \lambda_2 = tr A \\ \lambda_1 \lambda_2 = \det A. \end{cases} \tag{6.5}$$

Given the matrix of coefficients, A, in (6.1), it follows that

$$\begin{cases} \lambda_1 + \lambda_2 = 6.5 \\ \lambda_1 \lambda_2 = 3, \end{cases} \tag{6.6}$$

and therefore $\lambda_1 = 6$ and $\lambda_2 = 0.5$.

The Eigenvectors f_1 and f_2 of the Matrix A

The eigenvectors of the matrix A, f_1 and f_2, associated with the eigenvalues, λ_1 and λ_2, are obtained as solutions to the equations

$$\begin{aligned} [A - \lambda I] f_1 = 0 \quad for \quad f_1 \neq 0 \\ [A - \lambda I] f_2 = 0 \quad for \quad f_2 \neq 0, \end{aligned} \tag{6.7}$$

where $f_i = (f_{i1}, f_{i2})'$ for $i = 1, 2$. Hence, it follows from (6.1) that the eigenvector associated with the eigenvalue $\lambda_1 = 6$ is determined by the solution to the system of equations

$$\begin{bmatrix} -2 & 1 \\ 7 & -3.5 \end{bmatrix} \begin{bmatrix} f_{11} \\ f_{12} \end{bmatrix} = \begin{bmatrix} 0 \\ 0 \end{bmatrix}, \tag{6.8}$$

whereas that associated with $\lambda_2 = 0.5$ is determined by the solution to the system of equations

$$\begin{bmatrix} 3.5 & 1 \\ 7 & 2 \end{bmatrix} \begin{bmatrix} f_{21} \\ f_{22} \end{bmatrix} = \begin{bmatrix} 0 \\ 0 \end{bmatrix}. \tag{6.9}$$

Thus, the first eigenvector is determined by the equation

$$f_{12} = 2f_{11}, \tag{6.10}$$

whereas the second eigenvector is given by the equation

$$f_{21} = -3.5f_{22}. \tag{6.11}$$

The eigenvectors f_1 and f_2 are therefore

$$f_1 = \begin{bmatrix} 1 \\ 2 \end{bmatrix}$$

$$f_2 = \begin{bmatrix} 1 \\ -3.5 \end{bmatrix}, \tag{6.12}$$

or any scalar multiple of these vectors.

The Use of the Eigenvectors f_1 and f_2 in the Construction of a New System of Coordinates that Spans $\Re^2$

Since f_1 and f_2 are linearly independent, they span $\Re^2$. Namely, for all $x_t \in \Re^2$, there exists $y_t \equiv (y_{1t}, y_{2t}) \in \Re^2$ such that

$$x_t = f_1 y_{1t} + f_2 y_{2t}. \tag{6.13}$$

In other words, every $x_t \in \Re^2$ can be expressed in terms of the new system of coordinates (y_{1t}, y_{2t}). Hence, as follows from the values of the eigenvectors given in (6.12),

$$\begin{bmatrix} x_{1t} \\ x_{2t} \end{bmatrix} = \begin{bmatrix} 1 & 1 \\ 2 & -3.5 \end{bmatrix} \begin{bmatrix} y_{1t} \\ y_{2t} \end{bmatrix}. \tag{6.14}$$

Since, every $x_t = (x_{1t}, x_{2t})' \in \Re^2$ can be expressed in terms of the new system of coordinates, $(y_{1t}, y_{2t}) \in \Re^2$, there exists a time-independent matrix Q such that

$$x_t = Q y_t, \tag{6.15}$$

where

$$Q = \begin{bmatrix} 1 & 1 \\ 2 & -3.5 \end{bmatrix}. \tag{6.16}$$

Since f_1 and f_2 are linearly independent, Q is a non-singular matrix, Q^{-1} therefore exists, and y_t can be expressed in terms of the original system of coordinates, (x_{1t}, x_{2t}). That is,

$$y_t = Q^{-1} x_t. \tag{6.17}$$

In particular,

$$\begin{bmatrix} y_{1t} \\ y_{2t} \end{bmatrix} = -\frac{1}{5.5} \begin{bmatrix} -3.5 & -1 \\ -2 & 1 \end{bmatrix} \begin{bmatrix} x_{1t} \\ x_{2t} \end{bmatrix}, \tag{6.18}$$

and therefore,

$$\begin{aligned} y_{1t} &= \tfrac{1}{5.5}(3.5 x_{1t} + x_{2t}) \\ y_{2t} &= \tfrac{1}{5.5}(2 x_{1t} - x_{2t}). \end{aligned} \tag{6.19}$$

Thus,

$$\begin{aligned} y_{1t} = 0 &\Leftrightarrow x_{2t} = -3.5 x_{1t} \\ y_{2t} = 0 &\Leftrightarrow x_{2t} = 2 x_{1t}. \end{aligned} \tag{6.20}$$

The geometric place of the new system of coordinates is given by the set of two eqs. (6.20). As depicted in Fig. 6.1, the geometric place of all pairs (x_{1t}, x_{2t}) such that $x_{2t} = -3.5 x_{1t}$ is the y_{2t} axis (along which $y_{1t} = 0$), and the geometric place of all pairs (x_{1t}, x_{2t}) such that $x_{2t} = 2 x_{1t}$ is the y_{1t} axis (along which $y_{2t} = 0$).

The axes of the new system of coordinates (y_{1t}, y_{2t}) are therefore the lines spanned by the eigenvectors f_1 and f_2, respectively, as depicted in Fig. 6.1.

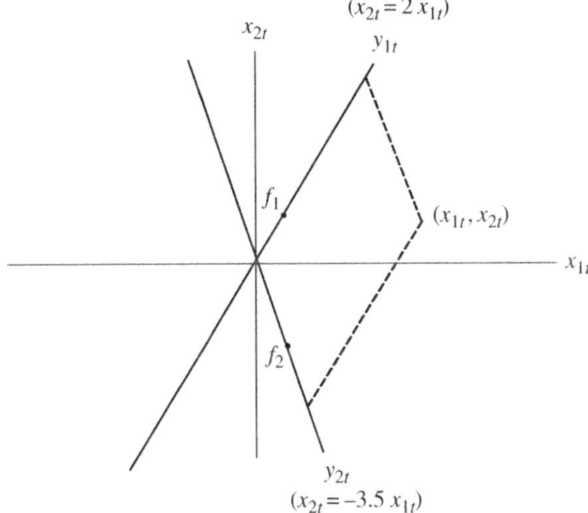

Fig. 6.1. The New System of Coordinates
The Representation of (x_{1t}, x_{2t}) in the (y_{1t}, y_{2t}) Space

The Time Path of the State Variables y_{1t} and y_{2t}

As follows from (6.17), the value of the vector of the new state variables, y_{t+1}, can be expressed as a time-invariant function of the value of the original vector of state variables, x_{t+1}. In particular,

$$y_{t+1} = Q^{-1}x_{t+1}. \tag{6.21}$$

Hence, since the evolution of the original vector of state variables x_{t+1} is given by $x_{t+1} = Ax_t$, it follows that

$$y_{t+1} = Q^{-1}Ax_t. \tag{6.22}$$

Moreover, as established in (6.15), the value of the original vector of state variables, x_t, can be expressed in terms of the new system of coordinates, (y_{1t}, y_{2t}). In particular, $x_t = Qy_t$, and therefore

$$y_{t+1} = Q^{-1}AQy_t. \tag{6.23}$$

Thus,

$$y_{t+1} \equiv Dy_t, \tag{6.24}$$

where $D \equiv Q^{-1}AQ$.

As follows from (6.15) and (6.18),

$$D = Q^{-1}AQ = -\frac{1}{5.5}\begin{bmatrix} -3.5 & -1 \\ -2 & 1 \end{bmatrix}\begin{bmatrix} 4 & 1 \\ 7 & 2.5 \end{bmatrix}\begin{bmatrix} 1 & 1 \\ 2 & -3.5 \end{bmatrix}$$

$$= \begin{bmatrix} 6 & 0 \\ 0 & 0.5 \end{bmatrix}. \tag{6.25}$$

Namely, the matrix D is a diagonal matrix whose elements are the eigenvalues of the matrix A, $\lambda_1 = 6$ and $\lambda_2 = 0.5$.

Thus, the evolution of each of the elements of the vector of the new state variables, y_t, is independent of the evolution of the other state variables, and its time path can be determined by the method of solution developed for the one-dimensional case in Sect. 1.1, as outlined below.

The evolution of the vector of state variables, y_t, is given therefore by

$$y_t = D^t y_0. \tag{6.26}$$

Namely,

$$y_{1t} = 6^t\, y_{10}$$
$$y_{2t} = 0.5^t\, y_{20}, \tag{6.27}$$

where the initial value of the vector of new state variables, y_0, is determined uniquely by the values of the vector of state variables in period 0, x_0. In particular, as follows from (6.17), $y_0 = Q^{-1}x_0$, i.e.

$$y_{10} = \tfrac{1}{5.5}\left(3.5x_{10} + x_{20}\right)$$
$$y_{20} = \tfrac{1}{5.5}\left(2x_{10} - x_{20}\right). \tag{6.28}$$

**The Stability of the Steady-State Equilibrium
of the System $y_{t+1} = Dy_t$**

The steady-state equilibrium of the system $y_{t+1} = Dy_t$ is a vector $\bar{y} \in \Re^2$ such that $\bar{y} = D\bar{y}$. The steady-state equilibrium of the new system is therefore

$$\bar{y} \equiv (\bar{y}_1, \bar{y}_2)' = (0,0)'. \tag{6.29}$$

The steady-state equilibrium $\bar{y} = (0,0)'$ is *unique* since $[I - D]$ is non-singular, i.e.

$$|I - D| = \begin{vmatrix} -5 & 0 \\ 0 & 0.5 \end{vmatrix} = -2.5 \neq 0. \tag{6.30}$$

The second state variable, y_{2t}, converges to its steady-state level, $\bar{y}_2 = 0$, regardless of its initial value, y_{20}. Namely,

$$\lim_{t \to \infty} y_{2t} = \lim_{t \to \infty} (0.5)^t \, y_{20} = \bar{y}_2 = 0, \quad \forall y_{20} \in \Re. \tag{6.31}$$

The first state variable, y_{1t}, diverges to plus or minus infinity, unless the initial position of this state variable is at its steady-state level, $\bar{y}_1 = 0$, i.e.

$$\lim_{t \to \infty} y_{1t} = \lim_{t \to \infty} 6^t \, y_{10} = \begin{cases} -\infty & if \quad y_{10} < 0 \\ \bar{y}_1 = 0 & if \quad y_{10} = 0 \\ \infty & if \quad y_{10} > 0. \end{cases} \tag{6.32}$$

As depicted in Fig. 6.2, the steady-state equilibrium $(\bar{y}_1, \bar{y}_2) = (0,0)$ is a saddle point. Namely, unless $y_{10} = 0$, the steady-state equilibrium will not be reached and the system will diverge in one of its dimensions to either plus or minus infinity:

$$\lim_{t \to \infty} y_t = \begin{cases} (-\infty, 0) & if \quad y_{10} < 0 \\ (0, 0) & if \quad y_{10} = 0 \\ (\infty, 0) & if \quad y_{10} > 0. \end{cases} \tag{6.33}$$

The Solution for x_t

The trajectory of the vector of original state variables, x_t, can be expressed in terms of the new system of coordinates, (y_{1t}, y_{2t}). In particular, as established in (6.15), $x_t = Q y_t$, i.e.,

$$\begin{bmatrix} x_{1t} \\ x_{2t} \end{bmatrix} = \begin{bmatrix} 1 & 1 \\ 2 & -3.5 \end{bmatrix} \begin{bmatrix} y_{1t} \\ y_{2t} \end{bmatrix}. \tag{6.34}$$

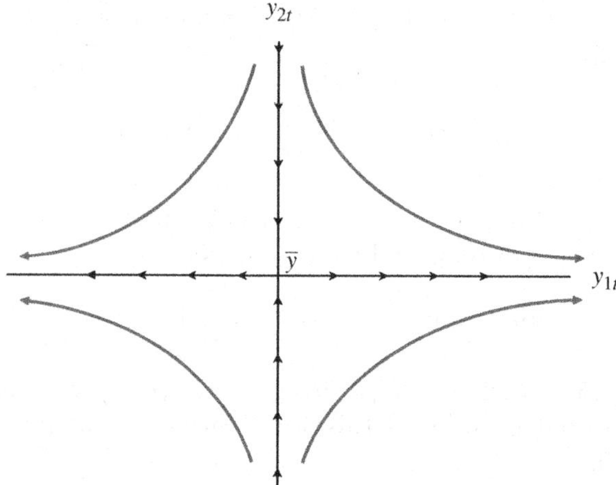

Fig. 6.2. The Evolution of y_t

Hence, a time-invariant transformation of the explicit solution for the time path of the vector of new state variables, y_t, provides an explicit solution for the time path of the original vector of state variables, x_t.

As follows from (6.27),

$$\begin{bmatrix} x_{1t} \\ x_{2t} \end{bmatrix} = \begin{bmatrix} 1 & 1 \\ 2 & -3.5 \end{bmatrix} \begin{bmatrix} 6^t y_{10} \\ (0.5)^t y_{20} \end{bmatrix} = \begin{bmatrix} 6^t y_{10} + (0.5)^t y_{20} \\ 2(6^t) y_{10} - 3.5(0.5)^t y_{20} \end{bmatrix}. \quad (6.35)$$

The time path of x_t and its qualitative properties are therefore uniquely determined by the system's initial conditions, (x_{10}, x_{20}), and the eigenvalues of the matrix A. As follows from (6.35), noting (6.28),

$$\begin{bmatrix} x_{1t} \\ x_{2t} \end{bmatrix} = \begin{bmatrix} \frac{1}{5.5} [6^t (3.5x_{10} + x_{20}) + (0.5)^t (2x_{10} - x_{20})] \\ \frac{1}{5.5} [2(6^t) (3.5x_{10} + x_{20}) - 3.5(0.5)^t (2x_{10} - x_{20})] \end{bmatrix}. \quad (6.36)$$

The phase diagram of the original system is obtained by placing the phase diagram that describes the evolution of y_t relative to the new system of coordinates (y_1, y_2), in the plane (x_1, x_2), as depicted in Fig. 6.3.

**The Stability of the Steady-State Equilibrium
of the System $x_{t+1} = Ax_t$**

A steady-state equilibrium of the system $x_{t+1} = Ax_t$ is a vector $\bar{x} \in \Re^2$ such that $\bar{x} = A\bar{x}$. Hence, it follows from (6.1) that $\bar{x}$ exists and

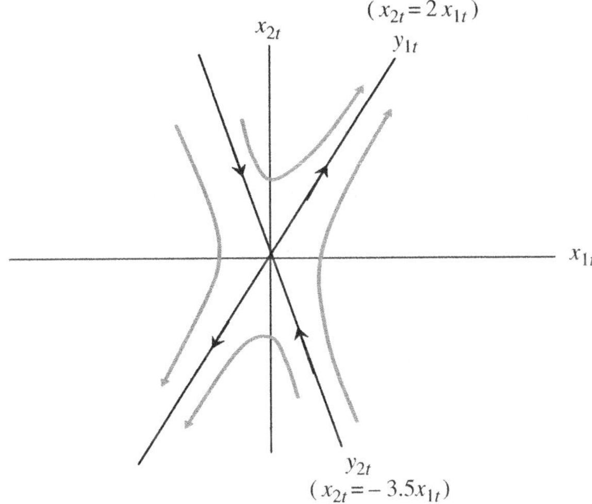

Fig. 6.3. The Evolution of x_t
A Saddle

is given by $\bar{x} = (0,0)$. Moreover, $\bar{x} = (0,0)$ is *unique* since the matrix $[I - A]$ is non-singular, i.e.

$$|I - A| = \begin{vmatrix} -3 & -1 \\ -7 & -1.5 \end{vmatrix} = -2.5 \neq 0. \tag{6.37}$$

As follows from (6.36), and as depicted in Fig. 6.3,[1]

$$\lim_{t \to \infty} x_t = \bar{x} \quad \Leftrightarrow \quad x_{20} = -3.5x_{10}, \tag{6.38}$$

and the steady-state equilibrium $\bar{x} = 0$ is a saddle point. Namely, the vector of state variables x_t converges to its steady-state value $\bar{x}$ if and only if the initial values of this vector are placed on the y_{2t} axis, i.e. if $x_{20} = -3.5x_{10}$.

B. Construction of a Phase Diagram Without an Explicit Solution

The derivation of a phase diagram for this two-dimensional, first-order linear system does not require an explicitcharacterization of the

---

[1] Since both eigenvalues are real and positive, convergence and divergence are monotonic and the arrows of motion approximate the actual motion of the system.

evolution of the vector of state variables. The phase diagram can be derived via a characterization of the map of forces that operate on the vector of state variables.

The construction of the phase diagram requires the identification of the geometric place under which each of the state variables is in a steady state and the characterization of the forces that operate on this state variable once it deviates from its steady-state value.

Let Δx_{it} be the change in the value of the ith state variable, $i = 1, 2$, from period t to period $t + 1$:

$$\Delta x_{1t} \equiv x_{1t+1} - x_{1t} = 3x_{1t} + x_{2t}$$
$$\Delta x_{2t} \equiv x_{2t+1} - x_{2t} = 7x_{1t} + 1.5x_{2t}.$$

$$(6.39)$$

Clearly, at a steady-state equilibrium, neither x_{1t} nor x_{2t} changes over time and therefore $\Delta x_{1t} = \Delta x_{2t} = 0$.

Let '$\Delta x_{1t} = 0$' be the geometric place of all pairs of (x_{1t}, x_{2t}) such that x_{1t} is in a steady state, and let '$\Delta x_{2t} = 0$' be the geometric place of all pairs (x_{1t}, x_{2t}) such that x_{2t} is in a steady state. Namely,

$$'\Delta x_{1t} = 0' \equiv \{(x_{1t}, x_{2t})|\ x_{1t+1} - x_{1t} = 0\}$$
$$'\Delta x_{2t} = 0' \equiv \{(x_{1t}, x_{2t})|\ x_{2t+1} - x_{2t} = 0\}.$$

$$(6.40)$$

It follows from (6.39) and (6.40) that

$$\Delta x_{1t} = 0 \quad \Leftrightarrow \quad x_{2t} = -3x_{1t}$$
$$\Delta x_{2t} = 0 \quad \Leftrightarrow \quad x_{2t} = -(14/3)x_{1t}.$$

$$(6.41)$$

Thus, as depicted in Fig. 6.4, the geometric locus '$\Delta x_{1t} = 0$' is given by the equation $x_{2t} = -3x_{1t}$, whereas the geometric locus '$\Delta x_{2t} = 0$' is given by the equation $x_{2t} = -(14/3)x_{1t}$. The geometric place under which the two loci, '$\Delta x_{1t} = 0$' and '$\Delta x_{2t} = 0$,' intersect is the steady-state equilibrium of the system. As follows from (6.41), and as depicted in Fig. 6.4, the two loci intersect at the point $(0, 0)$ and this is the unique steady-state equilibrium of the entire system.

The forces that operate on each of the state variables away from its steady-state equilibrium provide the necessary elements for the derivation of the phase diagram.

As follows from (6.39), as long as $x_{2t} > -3x_{1t}$, the system increases the value of the first state variable, x_{1t}, in the transition from time t

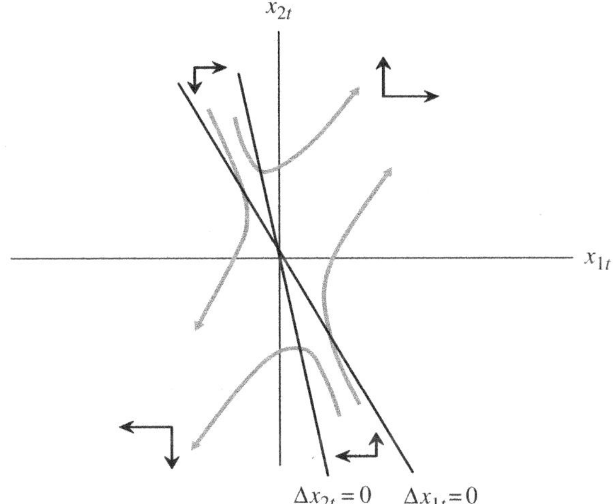

Fig. 6.4. The Phase Diagram without an Explicit Solution

to time $t+1$, whereas if $x_{2t} < -3x_{1t}$, the system decreases the value of x_{1t} in the transition from time t to time $t+1$. Hence,

$$\Delta x_{1t} \begin{cases} > 0 & if \quad x_{2t} > -3x_{1t} \\ \\ < 0 & if \quad x_{2t} < -3x_{1t}. \end{cases} \tag{6.42}$$

Consequently, as depicted in Fig. 6.4, above the line $x_{2t} = -3x_{1t}$ (i.e. for pairs (x_{1t}, x_{2t}) such that $x_{2t} > -3x_{1t}$), the arrows that depict the motion of the first state variable, x_{1t}, are directed rightward, whereas below the line (i.e. for pairs (x_{1t}, x_{2t}) such that $x_{2t} < -3x_{1t}$), the arrows that depict the motion of x_{1t} are directed leftward.

Similarly, if $x_{2t} > -(14/3)x_{1t}$, the system increases the value of the second state variable, x_{2t}, in the transition from time t to time $t+1$, whereas if $x_{2t} < -(14/3)x_{1t}$, the system decreases the value of x_{2t}. Hence,

$$\Delta x_{2t} \begin{cases} > 0 & if \quad x_{2t} > -(14/3)x_{1t} \\ \\ < 0 & if \quad x_{2t} < -(14/3)x_{1t}. \end{cases} \tag{6.43}$$

Consequently, as depicted in Fig. 6.4, above the geometric locus $x_{2t} = -(14/3)x_{1t}$ (i.e. for pairs (x_{1t}, x_{2t}) such that $x_{2t} > -(14/3)x_{1t}$), the arrows that depict the motion of the second state variable, x_{2t}, are directed upward, whereas below the geometric locus $x_{2t} = -(14/3)x_{1t}$ (i.e. for pairs (x_{1t}, x_{2t}) such that $x_{2t} < -(14/3)x_{1t}$), the arrows that depict the motion of x_{2t} are directed downward.

6.1.2 Complex Eigenvalues - Periodic Orbit

Consider the two-dimensional system of first-order homogenous linear difference equations

$$x_{t+1} = \begin{bmatrix} x_{1t+1} \\ x_{2t+1} \end{bmatrix} = \begin{bmatrix} \sqrt{2} & -1 \\ 1 & 0 \end{bmatrix} \begin{bmatrix} x_{1t} \\ x_{1t} \end{bmatrix} \equiv Ax_t, \tag{6.44}$$

where $x_0 \equiv [x_{10}, x_{20}]$ is given.

The characterization of the time path of this two-dimensional system of interdependent state variables is based on the construction of a time-independent transformation that converts the system into a new dynamical system whose evolution is governed by a matrix in the Jordan normal form.

The Eigenvalues μ and $\bar{\mu}$ of the Matrix of Coefficients A

The eigenvalues of the matrix A are obtained as a solution to the equation

$$|A - \mu I| = 0, \tag{6.45}$$

where $|A - \mu I|$ is the determinant of the matrix $[A - \mu I]$ and I is the identity matrix. The implied characteristic polynomial is therefore

$$c(\mu) \equiv \mu^2 - \sqrt{2}\mu + 1 = 0. \tag{6.46}$$

The eigenvalues of the coefficient matrix A are complex, taking the form $\mu \equiv \alpha + \beta i$ and $\bar{\mu} \equiv \alpha - \beta i$, where

$$\mu = \frac{\sqrt{2}}{2} + \frac{\sqrt{2}}{2}i$$

$$\bar{\mu} = \frac{\sqrt{2}}{2} - \frac{\sqrt{2}}{2}i, \tag{6.47}$$

and $i \equiv \sqrt{-1}$.

The Eigenvectors v and $\bar{v}$ of the Matrix of Coefficients A

The eigenvectors of the matrix A, v and $\bar{v}$, are obtained as a solution to the equations

$$[A - \mu I]v = 0 \quad for \quad \mu \neq 0$$
$$[A - \bar{\mu} I]\bar{v} = 0 \quad for \quad \bar{\mu} \neq 0. \tag{6.48}$$

Hence, the eigenvector v associated with the eigenvalue μ is determined by the solution to the system of equations

$$\begin{bmatrix} \frac{\sqrt{2}}{2} - \frac{\sqrt{2}}{2}i & -1 \\ 1 & -\frac{\sqrt{2}}{2} - \frac{\sqrt{2}}{2}i \end{bmatrix} \begin{bmatrix} v_1 \\ v_2 \end{bmatrix} = \begin{bmatrix} 0 \\ 0 \end{bmatrix}, \tag{6.49}$$

and the eigenvector $\bar{v}$ associated with the eigenvalue $\bar{\mu}$ is determined by the solution to the system of equations

$$\begin{bmatrix} \frac{\sqrt{2}}{2} + \frac{\sqrt{2}}{2}i & -1 \\ 1 & -\frac{\sqrt{2}}{2} + \frac{\sqrt{2}}{2}i \end{bmatrix} \begin{bmatrix} \bar{v}_1 \\ \bar{v}_2 \end{bmatrix} = \begin{bmatrix} 0 \\ 0 \end{bmatrix}. \tag{6.50}$$

Thus, the eigenvector v is determined by the equation

$$v_2 = \left(\frac{\sqrt{2}}{2} - \frac{\sqrt{2}}{2}i \right) v_1, \tag{6.51}$$

and the eigenvector $\bar{v}$ is determined by the equation

$$\bar{v}_2 = \left(\frac{\sqrt{2}}{2} + \frac{\sqrt{2}}{2}i \right) \bar{v}_1. \tag{6.52}$$

The eigenvectors v and $\bar{v}$ are therefore

$$v = \begin{bmatrix} 1 \\ \frac{\sqrt{2}}{2} - \frac{\sqrt{2}}{2}i \end{bmatrix}$$
$$\bar{v} = \begin{bmatrix} 1 \\ \frac{\sqrt{2}}{2} + \frac{\sqrt{2}}{2}i \end{bmatrix}, \tag{6.53}$$

or any scalar multiple of these vectors.

In particular, v and $\bar{v}$ can be decomposed into a real and complex part:

$$v \equiv u + wi = \begin{bmatrix} u_1 \\ u_2 \end{bmatrix} + \begin{bmatrix} w_1 \\ w_2 \end{bmatrix} i = \begin{bmatrix} 1 \\ \frac{\sqrt{2}}{2} \end{bmatrix} + \begin{bmatrix} 0 \\ -\frac{\sqrt{2}}{2} \end{bmatrix} i$$

$$\bar{v} \equiv u - wi = \begin{bmatrix} u_1 \\ u_2 \end{bmatrix} - \begin{bmatrix} w_1 \\ w_2 \end{bmatrix} i = \begin{bmatrix} 1 \\ \frac{\sqrt{2}}{2} \end{bmatrix} - \begin{bmatrix} 0 \\ -\frac{\sqrt{2}}{2} \end{bmatrix} i.$$

$$(6.54)$$

This decomposition is useful in the establishment of a new system of coordinates that spans $\Re^2$ and permits the transformation of the original system into a new one characterized by the Jordan normal form representation.

The Use of the Vectors u and w in the Construction of a New System of Coordinates that Spans $\Re^2$

Since u and w are linearly independent, they span $\Re^2$. Namely, for all $x_t \in \Re^2$ there exists $y_t \equiv (y_{1t}, y_{2t}) \in \Re^2$ such that

$$x_t = uy_{1t} - wy_{2t}. \tag{6.55}$$

In other words, every $x_t \in \Re^2$ can be expressed in terms of the new system of coordinates (y_{1t}, y_{2t}). As follows from the values of the eigenvectors given in (6.54),

$$\begin{bmatrix} x_{1t} \\ x_{2t} \end{bmatrix} = \begin{bmatrix} 1 & 0 \\ \frac{\sqrt{2}}{2} & \frac{\sqrt{2}}{2} \end{bmatrix} \begin{bmatrix} y_{1t} \\ y_{2t} \end{bmatrix}. \tag{6.56}$$

Namely, for all $x_t \in \Re^2$ there exists $y_t \equiv (y_{1t}, y_{2t}) \in \Re^2$ and a time-independent matrix V such that

$$x_t = Vy_t. \tag{6.57}$$

Since u and w are linearly independent, V is a non-singular matrix, V^{-1} therefore exists, and y_t can be expressed in terms of the original system of coordinates (x_{1t}, x_{2t}). Namely,

$$y_t = V^{-1}x_t, \tag{6.58}$$

i.e.

$$\begin{bmatrix} y_{1t} \\ y_{2t} \end{bmatrix} = \begin{bmatrix} 1 & 0 \\ -1 & \sqrt{2} \end{bmatrix} \begin{bmatrix} x_{1t} \\ x_{2t} \end{bmatrix}, \tag{6.59}$$

and therefore,

$$y_{1t} = x_{1t}$$
$$y_{2t} = \sqrt{2}x_{2t} - x_{1t}. \tag{6.60}$$

Thus,

$$y_{1t} = 0 \Leftrightarrow x_{1t} = 0$$
$$y_{2t} = 0 \Leftrightarrow x_{2t} = (1/\sqrt{2})x_{1t}. \tag{6.61}$$

The geometric place of the new system of coordinates is given by (6.61). As depicted in Fig. 6.6, the geometric place of all pairs (x_{1t}, x_{2t}) such that $x_{1t} = 0$ is the y_{2t} axis (along which $y_{1t} = 0$), and the geometric place of all pairs (x_{1t}, x_{2t}) such that $x_{2t} = (1/\sqrt{2})x_{1t}$ is the y_{1t} axis (along which $y_{2t} = 0$).

The Time Path of the State Variables y_{1t} and y_{2t}

As follows from (6.58), the vector of state variables, y_{t+1}, can be expressed as a time-invariant function of the vector of state variables, x_{t+1}. In particular,

$$y_{t+1} = V^{-1}x_{t+1}. \tag{6.62}$$

Hence, since the evolution of the vector of state variables, x_{t+1}, is given by $x_{t+1} = Ax_t$, it follows that

$$y_{t+1} = V^{-1}Ax_t. \tag{6.63}$$

Moreover, as established in (6.57), the vector of state variables, x_t, can be expressed in terms of the new system of coordinates, (y_{1t}, y_{2t}). In particular, $x_t = Vy_t$, and therefore

$$y_{t+1} = V^{-1}AVy_t. \tag{6.64}$$

Thus,

$$y_{t+1} \equiv D y_t, \tag{6.65}$$

where as follows from the (6.44), (6.56), and (6.59),

$$D = V^{-1} A V = \begin{bmatrix} \frac{\sqrt{2}}{2} & -\frac{\sqrt{2}}{2} \\ \frac{\sqrt{2}}{2} & \frac{\sqrt{2}}{2} \end{bmatrix}. \tag{6.66}$$

Hence,

$$y_{t+1} = \begin{bmatrix} y_{1t+1} \\ y_{2t+1} \end{bmatrix} = \begin{bmatrix} \frac{\sqrt{2}}{2} & -\frac{\sqrt{2}}{2} \\ \frac{\sqrt{2}}{2} & \frac{\sqrt{2}}{2} \end{bmatrix} \begin{bmatrix} y_{1t} \\ y_{2t} \end{bmatrix}, \tag{6.67}$$

where $y_0 \equiv (y_{10}, y_{20})' = v^{-1} x_0$ is given.

Let $\bar{y}$ be a steady-state equilibrium of the system, i.e.

$$\bar{y} = D\bar{y}. \tag{6.68}$$

Since $[I - D]$ is a non-singular matrix, the steady-state equilibrium, $\bar{y}$, is unique.

$$\bar{y} = (\bar{y}_1, \bar{y}_2) = (0, 0). \tag{6.69}$$

The trajectory $\{y_t\}_{t=0}^{\infty}$, satisfies (6.67) at all points in time and relates the value of the new vector of state variables at time t, y_t, to their initial condition, y_0, via the parameters embodied in the coefficient matrix A.

Following the method of iterations, the solution satisfies

$$\begin{bmatrix} y_{1t+1} \\ y_{2t+1} \end{bmatrix} = \begin{bmatrix} \frac{\sqrt{2}}{2} & -\frac{\sqrt{2}}{2} \\ \frac{\sqrt{2}}{2} & \frac{\sqrt{2}}{2} \end{bmatrix}^t \begin{bmatrix} y_{10} \\ y_{20} \end{bmatrix}. \tag{6.70}$$

As discussed in Sect. 3.3, this formulation is not very informative about the qualitative behavior of the dynamical system. In particular, it is not apparent what the necessary restrictions on the values of α and β are, such that the state variables will converge to their steady-state value. However, the evolution of each pair of state variables can be expressed in terms of the polar coordinates of (α, β).

Consider the geometrical representation of the complex pair of eigenvalues $\mu = \alpha + \beta i = \sqrt{2}/2 + \sqrt{2}/2i$ and $\bar{\mu} = \alpha - \beta i = \sqrt{2}/2 - \sqrt{2}/2i$ in the complex Cartesian space, as depicted in Fig. 3.12. Let r be the modulus of the eigenvalues, namely,

$$r \equiv \sqrt{\alpha^2 + \beta^2} = \sqrt{\left(\frac{\sqrt{2}}{2}\right)^2 + \left(\frac{\sqrt{2}}{2}\right)^2} = 1. \tag{6.71}$$

It follows that

$$\alpha = r\cos\theta$$
$$\beta = r\sin\theta, \tag{6.72}$$

and therefore

$$\begin{bmatrix} \alpha & -\beta \\ \beta & \alpha \end{bmatrix} = r \begin{bmatrix} \cos\theta & -\sin\theta \\ \sin\theta & \cos\theta \end{bmatrix}. \tag{6.73}$$

Hence, the analysis of the system in (6.70) can be expressed in terms of polar coordinates.

$$\begin{bmatrix} y_{1t} \\ y_{2t} \end{bmatrix} = \begin{bmatrix} \cos(\pi/4) & -\sin(\pi/4) \\ \sin(\pi/4) & \cos(\pi/4) \end{bmatrix}^t \begin{bmatrix} y_{10} \\ y_{20} \end{bmatrix}, \tag{6.74}$$

and therefore, noting Lemma 3.3,

$$\begin{bmatrix} y_{1t} \\ y_{2t} \end{bmatrix} = \begin{bmatrix} \cos(t\pi/4) & -\sin(t\pi/4) \\ \sin(t\pi/4) & \cos(t\pi/4) \end{bmatrix} \begin{bmatrix} y_{10} \\ y_{20} \end{bmatrix}. \tag{6.75}$$

Hence the trajectory of y_t is given by

$$y_{1t} = y_{10}\cos(t\pi/4) - y_{20}\sin(t\pi/4)$$
$$y_{2t} = y_{10}\sin(t\pi/4) + y_{20}\cos(t\pi/4), \tag{6.76}$$

where

$$\lim_{t\to\infty} y_t = \bar{y} = 0 \Leftrightarrow y_0 = \bar{y} = 0. \tag{6.77}$$

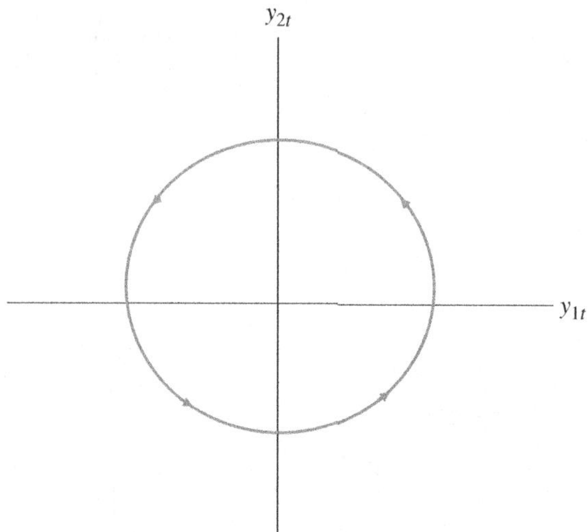

Fig. 6.5. The Evolution of y_t
(Counter-Clockwise) Periodic Orbit

Thus, as follows from (6.76), all trajectories of y_t initiated from any non-zero initial condition exhibit counter-clockwise periodic orbits, as depicted in Fig. 6.5. A full revolution of the orbit is completed every 8 periods.

The Time Path of x_t

The trajectory of the vector of state variables, x_t, can be expressed in terms of the trajectory of y_t. In particular, as established in (6.56), $x_t = V y_t$, i.e.

$$\begin{bmatrix} x_{1t} \\ x_{2t} \end{bmatrix} = \begin{bmatrix} 1 & 0 \\ \frac{\sqrt{2}}{2} & \frac{\sqrt{2}}{2} \end{bmatrix} \begin{bmatrix} y_{1t} \\ y_{2t} \end{bmatrix}. \tag{6.78}$$

Hence, a time-invariant transformation of the explicit solution for the time path of the vector of new state variables, y_t, provides an explicit solution for the time path of the original vector of state variables, x_t.

The time path of x_t and its qualitative properties are therefore uniquely determined by the system's initial conditions, (x_{10}, x_{20}), and the modulus of the eigenvalues of the matrix A. As follows from (6.76) and (6.78),

$$
\begin{bmatrix} x_{1t} \\ x_{2t} \end{bmatrix} = \begin{bmatrix} 1 & 0 \\ \dfrac{\sqrt{2}}{2} & \dfrac{\sqrt{2}}{2} \end{bmatrix} \begin{bmatrix} y_{10}\cos(t\pi/4) - y_{20}\sin(t\pi/4) \\ y_{10}\sin(t\pi/4) + y_{20}\cos(t\pi/4) \end{bmatrix}, \tag{6.79}
$$

and therefore

$$
\begin{bmatrix} x_{1t} \\ x_{2t} \end{bmatrix} = \begin{bmatrix} y_{10}\cos(t\pi/4) - y_{20}\sin(t\pi/4) \\ \dfrac{\sqrt{2}}{2}\left[(y_{10}+y_{20})\cos(t\pi/4) + (y_{10}-y_{20})\sin(t\pi/4)\right] \end{bmatrix}. \tag{6.80}
$$

Noting that $y_0 = v^{-1}x_0$, it follows from (6.60) that

$$
\begin{bmatrix} x_{1t} \\ x_{2t} \end{bmatrix} = \begin{bmatrix} x_{10}\cos(t\pi/4) - (\sqrt{2}x_{20} - x_{10})\sin(t\pi/4) \\ x_{20}[\cos(t\pi/4) - \sin(t\pi/4)] + \sqrt{2}x_{10}\sin(t\pi/4) \end{bmatrix}. \tag{6.81}
$$

The phase diagram of the system is obtained by placing the phase diagram that describes the evolution of y_t relative to the new system of coordinates (y_1, y_2), in the plane (x_1, x_2). As depicted in Fig. 6.6, the trajectory of x_t from any non-zero initial condition exhibits counter-clockwise periodic orbits.

The qualitative trajectory of the state variables could be inferred directly from the modulus, r, of these complex eigenvalues. As discussed in Sect. 3.3 and defined in (6.71), the modulus, r, of the eigenvalues of the coefficient matrix A is equal to one. Hence, as implied by Theorem 3.4, the system is characterized by a counter-clockwise ($\beta > 0$) periodic orbit.

6.1.3 Complex Eigenvalues - Spiral Sink

Consider the two-dimensional system of first-order homogenous linear difference equations.

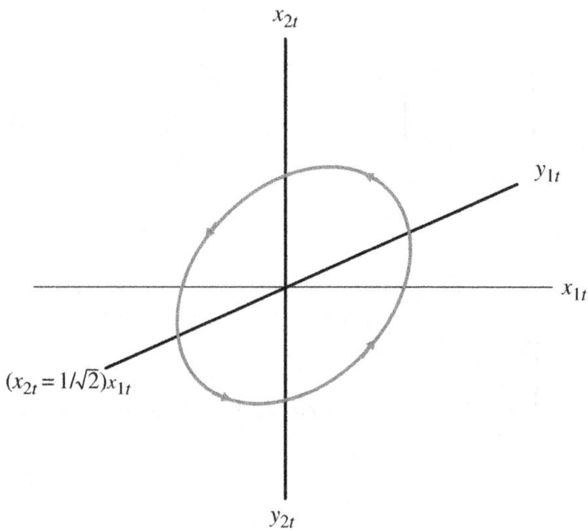

Fig. 6.6. The Evolution of x_t
(Counter-Clockwise) Periodic Orbit

$$y_{t+1} = \begin{bmatrix} y_{1t+1} \\ y_{2t+1} \end{bmatrix} = \begin{bmatrix} \dfrac{1}{2} & -\dfrac{1}{2} \\ \dfrac{1}{2} & \dfrac{1}{2} \end{bmatrix} \begin{bmatrix} y_{1t} \\ y_{2t} \end{bmatrix} \equiv Ay_t, \qquad (6.82)$$

where $y_0 \equiv [y_{10}, y_{20}]$ is given.

The eigenvalues of the coefficient matrix A are complex, taking the form $\mu \equiv \alpha + \beta i$ and $\overline{\mu} \equiv \alpha - \beta i$, where

$$\mu = \frac{1}{2} + \frac{1}{2}i,$$

$$\overline{\mu} = \frac{1}{2} - \frac{1}{2}i, \qquad (6.83)$$

and the coefficient matrix A is therefore already in Jordan normal form.

Let $\bar{y}$ be a steady-state equilibrium of the system, i.e.

$$\bar{y} = A\bar{y}. \qquad (6.84)$$

Since $[I - A]$ is a non-singular matrix, the steady-state equilibrium, $\bar{y}$, is unique.

$$\bar{y} = (\bar{y}_1, \bar{y}_2) = (0,0). \qquad (6.85)$$

The solution to this system is a trajectory, $\{y_t\}_{t=0}^{\infty}$, that satisfies (6.82) at all points in time and relates the value of the vector of state variables at time t, y_t, to their initial condition, y_0, via the parameters embodied in the coefficient matrix A. Following the method of iterations, the solution satisfies

$$\begin{bmatrix} y_{1t} \\ y_{2t} \end{bmatrix} = \begin{bmatrix} \frac{1}{2} & -\frac{1}{2} \\ \frac{1}{2} & \frac{1}{2} \end{bmatrix}^t \begin{bmatrix} y_{10} \\ y_{20} \end{bmatrix}. \tag{6.86}$$

As discussed in Sect. 3.3, this formulation is not very informative about the qualitative behavior of the dynamical system. In particular, it is not apparent what the necessary restrictions on the values of α and β are, such that the state variables will converge to their steady-state value. However, the evolution of each pair of state variables can be expressed in terms of the polar coordinates of (α, β).

Consider the geometrical representation of the complex pair of eigenvalues, $\mu = \alpha + \beta i = 0.5 + 0.5i$ and $\bar{\mu} = \alpha - \beta i = 0.5 - 0.5i$, in the complex Cartesian space, as depicted in Fig. 3.12. Let r be the modulus of the eigenvalues, namely,

$$r \equiv \sqrt{\alpha^2 + \beta^2} = \sqrt{\left(\frac{1}{2}\right)^2 + \left(\frac{1}{2}\right)^2} < 1. \tag{6.87}$$

It follows that

$$\alpha = r\cos\theta$$
$$\beta = r\sin\theta, \tag{6.88}$$

and therefore

$$\begin{bmatrix} \alpha & -\beta \\ \beta & \alpha \end{bmatrix} = r \begin{bmatrix} \cos\theta & -\sin\theta \\ \sin\theta & \cos\theta \end{bmatrix}. \tag{6.89}$$

Hence, the analysis of the system in (6.82) can be expressed in terms of polar coordinates:

$$\begin{bmatrix} y_{1t} \\ y_{2t} \end{bmatrix} = r^t \begin{bmatrix} \cos(\pi/4) & -\sin(\pi/4) \\ \sin(\pi/4) & \cos(\pi/4) \end{bmatrix}^t \begin{bmatrix} y_{10} \\ y_{20} \end{bmatrix}, \tag{6.90}$$

where $r = 1/\sqrt{2}$. Following Lemma 3.3,

$$\begin{bmatrix} y_{1t} \\ y_{2t} \end{bmatrix} = \begin{bmatrix} \dfrac{1}{\sqrt{2}} \end{bmatrix}^{t} \begin{bmatrix} \cos(t\pi/4) & -\sin(t\pi/4) \\ \sin(t\pi/4) & \cos(t\pi/4) \end{bmatrix} \begin{bmatrix} y_{10} \\ y_{20} \end{bmatrix}, \tag{6.91}$$

and the trajectory of y_t is given by

$$y_{1t} = \left[\frac{1}{\sqrt{2}}\right]^{t} [y_{10}\cos(t\pi/4) - y_{20}\sin(t\pi/4)]$$

$$\tag{6.92}$$

$$y_{2t} = \left[\frac{1}{\sqrt{2}}\right]^{t} [y_{10}\sin(t\pi/4) + y_{20}\cos(t\pi/4)].$$

Since $\cos(t\pi/4)$ and $\sin(t\pi/4)$ are bounded by in the interval $[-1, 1]$, it follows that

$$\lim_{t\to\infty} y_t = \bar{y} = 0 \qquad \forall y_0 \in \Re^2. \tag{6.93}$$

The steady-state equilibrium, $\bar{y}$, is globally stable. All trajectories that are initiated from a non-zero initial condition exhibit counter-clockwise spiral sink, as depicted in Fig. 6.7.

The qualitative trajectory of the state variable could be inferred directly from the modulus, r, of these complex eigenvalues. As discussed in Sect. 3.3 and defined in (6.87), the modulus, r, of the eigenvalues of the coefficient matrix A is smaller than one. Hence, as followed from Theorem 3.4, the system is characterized by a counter-clockwise ($\beta > 0$) spiral sink.

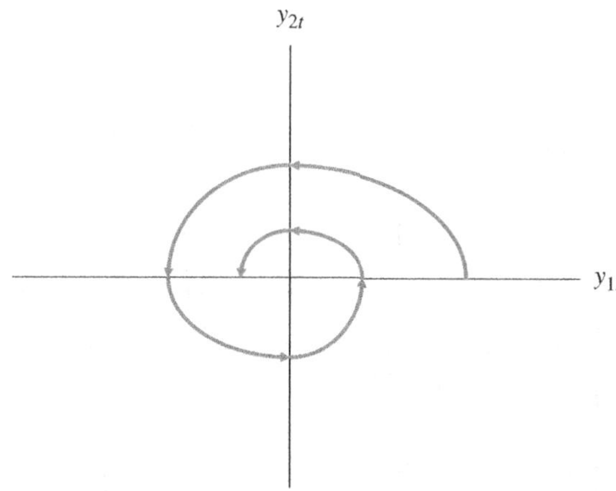

Fig. 6.7. (Counter-Clockwise) Spiral Sink

6.2 Second-Order Linear Systems

Consider a one-dimensional, *second-order*, autonomous, linear difference equation that governs the evolution of a one-dimensional state variable, x_{1t}, over time. Unlike the first-order case, the value of the state variable at time $t + 2$ depends on its value at time $t + 1$ and at time t. In particular, consider the difference equation

$$x_{1t+2} = -0.5x_{1t+1} + 1.5x_{1t}, \tag{6.94}$$

where the initial value of the state variable at time 0, x_{10}, and time $1, x_{11}$, are given.

In order to examine this difference equation in a familiar manner, using the basic propositions established in Chaps. 2–4, this one-dimensional second-order difference equation is converted into a two-dimensional first-order system.

Define a new state variable, x_{2t}, such that,

$$x_{1t+1} \equiv x_{2t}, \tag{6.95}$$

and therefore

$$x_{1t+2} = x_{2t+1}. \tag{6.96}$$

Substituting (6.95) and (6.96) into (6.94), the one-dimensional, second-order difference equation can be transformed into the two-dimensional first-order system,

$$\begin{aligned}
x_{1t+1} &= x_{2t} \\
x_{2t+1} &= 1.5x_{1t} - 0.5x_{2t}.
\end{aligned} \tag{6.97}$$

Hence, the one-dimensional, second-order difference equation is converted into a two-dimensional first-order system that characterizes the evolution of two state variables, $\{x_{2t}, x_{1t}\}$, over time:

$$\begin{bmatrix} x_{1t+1} \\ x_{2t+1} \end{bmatrix} = \begin{bmatrix} 0 & 1 \\ 1.5 & -0.5 \end{bmatrix} \begin{bmatrix} x_{1t} \\ x_{2t} \end{bmatrix}, \tag{6.98}$$

where the initial conditions of the two state variables, (x_{10}, x_{20}), are given since (x_{10}, x_{11}) are given, and as follow from (6.95), $x_{20} = x_{11}$.

The characterization of the time path of this two-dimensional system of interdependent state variables is based on the construction of a time-independent transformation that converts the system into a new dynamical system of independent state variables whose evolution can be derived based on the analysis of the one-dimensional case.

Steady-State Equilibria

Let $(\bar{x}_1, \bar{x}_2)$ be a steady-state equilibrium of the system. It follows from (6.98), or alternatively from (6.94), that there exists a continuum of steady-state equilibria defined by the 45^0 line in the plane (x_{1t}, x_{2t}). Namely,

$$\{(x_{1t}, x_{2t}) \in \Re^2 : (x_{1t}, x_{2t}) \\ = (\bar{x}_1, \bar{x}_2)\} = \{(x_{1t}, x_{2t}) \in \Re^2 : x_{2t} = x_{1t}\}. \tag{6.99}$$

The Eigenvalues λ_1 and λ_2 of the Matrix A

As established in Sect. 6.1.1, the eigenvalues (λ_1, λ_2) of the coefficient matrix A are given by the solution to

$$\begin{cases} \lambda_1 + \lambda_2 = & trA = -0.5 \\ \\ \lambda_1 \lambda_2 & = \det A = -1.5. \end{cases} \tag{6.100}$$

Hence,[2]

$$(\lambda_1, \lambda_2) = (1, -1.5).$$

The Eigenvectors f_1 and f_2 of the Matrix A

The eigenvector of the matrix A, f_1 and f_2, associated with the eigenvalues λ_1 and λ_2, are obtained as a solution to the equations

$$[A - \lambda I]f_1 = 0 \quad for \quad f_1 \neq 0$$

$$[A - \lambda I]f_2 = 0 \quad for \quad f_2 \neq 0, \tag{6.101}$$

where $f_i = (f_{i1}, f_{i2})'$ for $i = 1, 2$. Hence, it follows from (6.98) that the eigenvector associated with the eigenvalue $\lambda_1 = 1$ is determined by the solution to the system of equations

---

[2] It should be noted that in other examples of second-order, one-dimensional systems, the eigenvalues of the matrix of coefficients of the two-dimensional, first-order system, to which the second-order system was transformed, need not be equal to one.

$$\begin{bmatrix} -1 & 1 \\ 1.5 & -1.5 \end{bmatrix} \begin{bmatrix} f_{11} \\ f_{12} \end{bmatrix} = \begin{bmatrix} 0 \\ 0 \end{bmatrix}, \tag{6.102}$$

whereas that associated with $\lambda_2 = -1.5$ is determined by the solution to the system of equations

$$\begin{bmatrix} 1.5 & 1 \\ 1.5 & 1 \end{bmatrix} \begin{bmatrix} f_{21} \\ f_{22} \end{bmatrix} = 0. \tag{6.103}$$

Thus, the first eigenvector is determined by the equation

$$f_{12} = f_{11}, \tag{6.104}$$

and the second eigenvector is determined by the equation

$$f_{21} = -1.5 f_{22}. \tag{6.105}$$

The eigenvectors f_1 and f_2 are therefore

$$f_1 = \begin{bmatrix} 1 \\ 1 \end{bmatrix}$$

$$f_2 = \begin{bmatrix} 1 \\ -1.5 \end{bmatrix}, \tag{6.106}$$

or any scalar multiple of these vectors.

The Use of the Eigenvectors f_1 and f_2 in the Construction of a New System of Coordinates that Spans $\Re^2$

Since f_1 and f_2 are linearly independent, they span $\Re^2$. Namely, for every $x_t \in \Re^2$ there exists $y_t \equiv (y_{1t}, y_{2t}) \in \Re^2$ such that

$$x_t = f_1 y_{1t} + f_2 y_{2t}. \tag{6.107}$$

In other words, every $x_t \in \Re^2$ can be expressed in terms of the new system of coordinates, (y_1, y_2). Hence, as follows from the values of the eigenvectors given in (6.106),

$$\begin{bmatrix} x_{1t} \\ x_{2t} \end{bmatrix} = \begin{bmatrix} 1 & 1 \\ 1 & -1.5 \end{bmatrix} \begin{bmatrix} y_{1t} \\ y_{2t} \end{bmatrix}. \tag{6.108}$$

Namely, every $x_t = (x_{1t}, x_{2t})' \in \Re^2$ can be expressed in terms of the new system of coordinates, $(y_{1t}, y_{2t}) \in \Re^2$. Thus, there exists a time-independent matrix Q such that

$$x_t = Qy_t. \tag{6.109}$$

Since f_1 and f_2 are linearly independent, Q is a non-singular matrix, Q^{-1} therefore exists, and y_t can be expressed in terms of the original system of coordinates, (x_{1t}, x_{2t}). That is,

$$y_t = Q^{-1}x_t. \tag{6.110}$$

In particular,

$$\begin{bmatrix} y_{1t} \\ y_{2t} \end{bmatrix} = -0.4 \begin{bmatrix} -1.5 & -1 \\ -1 & 1 \end{bmatrix} \begin{bmatrix} x_{1t} \\ x_{2t} \end{bmatrix}, \tag{6.111}$$

and therefore,

$$y_{1t} = 0.6x_{1t} + 0.4x_{2t}$$
$$y_{2t} = 0.4(x_{1t} - x_{2t}). \tag{6.112}$$

Thus,

$$y_{1t} = 0 \Leftrightarrow x_{2t} = -1.5x_{1t}$$
$$y_{2t} = 0 \Leftrightarrow x_{2t} = x_{1t}. \tag{6.113}$$

The geometric place of the new system of coordinates is given by (6.113). As depicted in Fig. 6.8, the geometric place of all pairs (x_{1t}, x_{2t}) such that $x_{2t} = -1.5x_{1t}$ is the y_{2t} axis (along which $y_{1t} = 0$), and the geometric place of all pairs (x_{1t}, x_{2t}) such that $x_{2t} = x_{1t}$ is the y_{1t} axis (along which $y_{2t} = 0$). The axes of the new system of coordinates, (y_{1t}, y_{2t}), are therefore the lines spanned by the eigenvectors f_1 and f_2, respectively, as depicted in Fig. 6.8.

The Time Path of the State Variables y_{1t} and y_{2t}

As follows from (6.110), the value of the vector of state variables, y_{t+1}, can be expressed as a time-invariant function of the value of the original vector of state variables, x_{t+1}. In particular,

$$y_{t+1} = Q^{-1}x_{t+1}. \tag{6.114}$$

Hence, since the evolution of the vector of state variables, x_{t+1}, is given by $x_{t+1} = Ax_t$, it follows that

$$y_{t+1} = Q^{-1}Ax_t. \tag{6.115}$$

Moreover, as established in (6.109), the value of the vector of state variables x_t can be expressed in terms of the new system of coordinates, (y_{1t}, y_{2t}). In particular, $x_t = Qy_t$, and therefore $y_{t+1} = Q^{-1}AQy_t$. Thus,

$$y_{t+1} \equiv Dy_t, \tag{6.116}$$

where

$$D = Q^{-1}AQ = \begin{bmatrix} 1 & 0 \\ 0 & -1.5 \end{bmatrix}. \tag{6.117}$$

Thus, the evolution of each of the elements of the new vector of state variables, y_t, is independent of the evolution of the other state variable, and its time path can be determined by the method of solution developed for the one-dimensional case in Sect. 1.1.

The evolution of the new vector of state variables, y_t, is given by

$$y_t = D^t y_0, \tag{6.118}$$

and therefore

$$\begin{aligned} y_{1t} &= y_{10} \\ y_{2t} &= (-1.5)^t \, y_{20}. \end{aligned} \tag{6.119}$$

The value of the vector of new state variables in period 0, y_0, is determined uniquely by the values of the vector of the original state variables in period 0, x_0. In particular, as follows from (6.113),

$$\begin{aligned} y_{10} &= 0.6x_{10} + 0.4x_{20} \\ y_{20} &= 0.4\left(x_{10} - x_{20}\right). \end{aligned} \tag{6.120}$$

Instability of the Steady-State Equilibrium of the System $y_{t+1} = Dy_t$

The steady-state equilibrium of the system $y_{t+1} = Dy_t$ is a vector $\bar{y} \in \Re^2$ such that $\bar{y} = D\bar{y}$. The steady-state equilibrium of the new system is therefore

$$\bar{y} \equiv (\bar{y}_1, \bar{y}_2) = (0, 0). \tag{6.121}$$

Moreover, $\bar{y} = (0, 0)$ is *unique* since $[I - D]$ is non-singular.

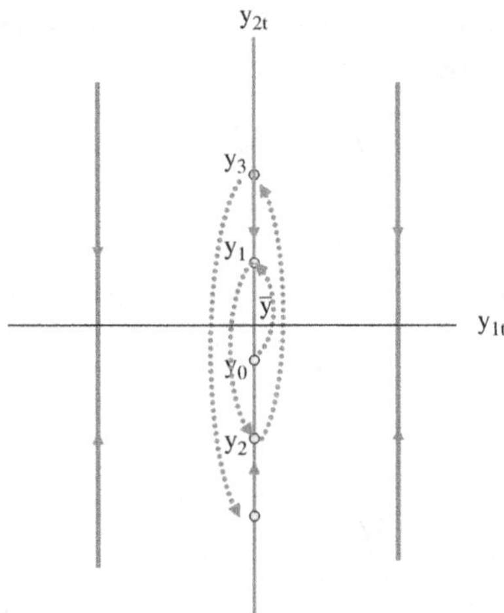

Fig. 6.8. (Vertical) Oscillatory Divergence

As depicted in Fig. 6.8, the dynamical system is characterized by (vertical) oscillatory divergence.

In particular, the second state variable, y_{2t}, diverges in oscillations as long as its initial value, y_{20}, is different from $\bar{y}_2$, i.e.

$$\lim_{t \to \infty} |y_{2t}| = \infty, \quad \forall y_{20} \neq \bar{y}_2, \tag{6.122}$$

whereas the first state variable, y_{1t}, remains at its initial value, y_{10}.

The Solution for x_t

The trajectory of the original vector of state variables, x_t, can be expressed in terms, of the new system of coordinates, (y_{1t}, y_{2t}). In particular, as established in (6.109), $x_t = Qy_t$, and thus it follows from (6.108) that

$$\begin{bmatrix} x_{1t} \\ x_{2t} \end{bmatrix} = \begin{bmatrix} 1 & 1 \\ 1 & -1.5 \end{bmatrix} \begin{bmatrix} y_{10} \\ (-1.5)^t y_{20} \end{bmatrix} = \begin{bmatrix} y_{10} + (-1.5)^t y_{20} \\ y_{10} + (-1.5)^{t+1} y_{20} \end{bmatrix}. \tag{6.123}$$

The time path of x_t and its qualitative properties are therefore uniquely determined by the system's initial conditions, (x_{10}, x_{20}), and the eigenvalues of the matrix A. As follows from (6.120) and (6.123),

$$\begin{bmatrix} x_{1t} \\ x_{2t} \end{bmatrix} = \begin{bmatrix} 0.6x_{10} + 0.4x_{20} + 0.4(-1.5)^t\,(x_{10} - x_{20}) \\ 0.6x_{10} + 0.4x_{20} + 0.4(-1.5)^{t+1}\,(x_{10} - x_{20}) \end{bmatrix}. \tag{6.124}$$

The phase diagram of the original system is obtained by placing the phase diagram that describes the evolution of y_t relative to the new system of coordinates (y_1, y_2), in the plane (x_1, x_2), as depicted in Fig. 6.9.

The Instability of Steady-State Equilibria of the System $x_{t+1} = Ax_t$

A steady-state equilibrium of the system $x_{t+1} = Ax_t$ is a vector $\bar{x} \in \Re^2$ such that $\bar{x} = A\bar{x}$. Hence, it follows from (6.98) that there exists a continuum of steady-state equilibria defined by the 45^0 line (i.e., $x_{2t} = x_{1t}$) in the plane (x_{1t}, x_{2t}).

As follows from (6.124), and as depicted in Fig. 6.9,

$$\lim_{t \to \infty} x_t = \bar{x} \quad \Leftrightarrow \quad x_{20} = x_{10}, \tag{6.125}$$

and all steady-state equilibria are unstable.

Namely, the vector of the original state variables, x_t, will be at a steady-state value $\bar{x}$ if and only if the initial values of this vector

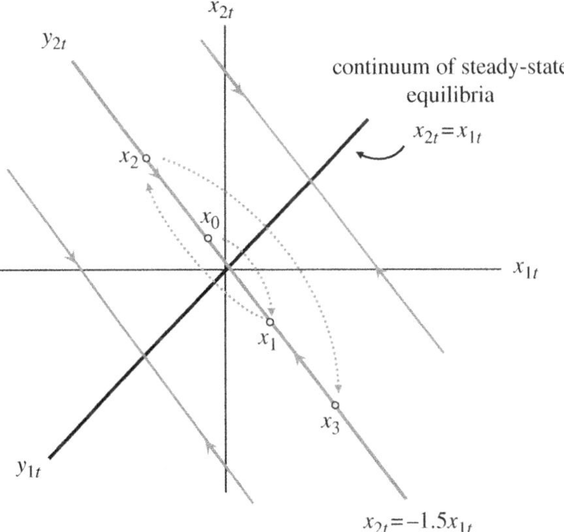

Fig. 6.9. The Evolution of x_t
Oscillatory Divergence

are such that $x_{20} = x_{10}$. Otherwise the state variable diverges in oscillations, as depicted in Fig. 6.9.

6.3 Nonlinear Systems

Consider a two-dimensional system of nonlinear first-order difference equations

$$
\begin{aligned}
x_{1t+1} &= \phi^1(x_{1t}, x_{2t}) = \sqrt{(2 - x_{1t})} \\
x_{2t+1} &= \phi^2(x_{1t}, x_{2t}) = 0.5(x_{1t} + \sqrt{x_{2t}}).
\end{aligned}
\tag{6.126}
$$

A steady-state equilibrium of the system, $\bar{x} \equiv (\bar{x}_1, \bar{x}_2)$, requires that

$$
\begin{aligned}
\bar{x}_1 &= \phi^1(\bar{x}_1, \bar{x}_2) = \sqrt{(2 - \bar{x}_1)} \\
\bar{x}_2 &= \phi^2(\bar{x}_1, \bar{x}_2) = 0.5(\bar{x}_1 + \sqrt{\bar{x}_2}).
\end{aligned}
\tag{6.127}
$$

Hence $(\bar{x}_1, \bar{x}_2) = (1, 1)$ is a steady-state equilibrium.

As established in (4.5), a nonlinear system can be approximated locally around the steady-state equilibrium by the linear system. Thus,

$$
\begin{bmatrix} x_{1t+1} \\ x_{2t+1} \end{bmatrix} =
\begin{bmatrix} \dfrac{\partial \phi^1(\bar{x})}{\partial x_{1t}} & \dfrac{\partial \phi^1(\bar{x})}{\partial x_{2t}} \\ \dfrac{\partial \phi^2(\bar{x})}{\partial x_{1t}} & \dfrac{\partial \phi^2(\bar{x})}{\partial x_{2t}} \end{bmatrix}
\begin{bmatrix} x_{1t} \\ x_{2t} \end{bmatrix} +
\begin{bmatrix} \phi^1(\bar{x}) - \sum_{j=1}^{2} \dfrac{\partial \phi^1(\bar{x})}{\partial x_{jt}}\bar{x}_j \\ \phi^2(\bar{x}) - \sum_{j=1}^{2} \dfrac{\partial \phi^2(\bar{x})}{\partial x_{jt}}\bar{x}_j \end{bmatrix},
\tag{6.128}
$$

where as follows from (6.126),

$$
\begin{bmatrix} \dfrac{\partial \phi^1(\bar{x})}{\partial x_{1t}} & \dfrac{\partial \phi^1(\bar{x})}{\partial x_{2t}} \\ \dfrac{\partial \phi^2(\bar{x})}{\partial x_{1t}} & \dfrac{\partial \phi^2(\bar{x})}{\partial x_{2t}} \end{bmatrix} =
\begin{bmatrix} -\dfrac{1}{2\sqrt{2 - \bar{x}_1}} & 0 \\ \dfrac{1}{2} & \dfrac{1}{4\sqrt{\bar{x}_2}} \end{bmatrix} =
\begin{bmatrix} -\dfrac{1}{2} & 0 \\ \dfrac{1}{2} & \dfrac{1}{4} \end{bmatrix}.
\tag{6.129}
$$

The linearized system around $(\bar{x}_1, \bar{x}_2) = (1, 1)$ is therefore

$$\begin{bmatrix} x_{1t+1} \\ x_{2t+1} \end{bmatrix} = \begin{bmatrix} -\frac{1}{2} & 0 \\ \frac{1}{2} & \frac{1}{4} \end{bmatrix} \begin{bmatrix} x_{1t} \\ x_{2t} \end{bmatrix} + \begin{bmatrix} \frac{3}{2} \\ \frac{1}{4} \end{bmatrix}. \tag{6.130}$$

The Jacobian matrix of coefficients is

$$J(1,1) \equiv \begin{bmatrix} -\frac{1}{2} & 0 \\ \frac{1}{2} & \frac{1}{4} \end{bmatrix}, \tag{6.131}$$

and the eigenvalues of $J(1,1)$ are therefore

$$\lambda_1 = -\frac{1}{2}$$
$$\lambda_2 = \frac{1}{4}. \tag{6.132}$$

Since both eigenvalues are smaller than 1 in absolute value and one of them is negative, the nonlinear dynamical system is locally stable in the neighborhood of the steady-state equilibrium $(\bar{x}_1, \bar{x}_2) = (1,1)$ and convergence is oscillatory.

Glossary

N	set of natural numbers
$\Re$	set of real numbers
$\Re_+$	set of non-negative real numbers
$\Re^n$	n-dimensional Euclidian space
$x \in \Re$	x is a real number
$x \in \Re^n$	x is an n-dimensional real vector
$\forall x$	for all x
$\exists x$	for some x
$|x|$	Euclidian norm of x
$\{x_i\}_{i=0}^{\infty}$	infinite sequence
$tr A$	trace of the matrix A
$|A|$	determinant of the matrix A
$\det A$	determinant of the matrix A
$A \Leftrightarrow B$	A if and only if B
λ	real eigenvalue
μ	complex eigenvalue
f	real eigenvector
v	complex eigenvector
$c(\lambda)$	characteristic polynomial

(a, b)	open interval
$[a, b]$	closed interval
(x, y)	ordered pair
$\{x, y\}$	unordered pair
$B_\epsilon(x)$	open ball of radius ϵ around x
$D\phi(x)$	total derivative of $\phi(x)$ with respect to x
$\partial\phi(x)/\partial x_i$	partial derivative of $\phi(x)$ with respect to x_i
$\phi^{\{n\}}(x)$	n^{th} forward iteration over x under the map ϕ
$\phi^{-\{n\}}(x)$	n^{th} backward iteration over x under the map ϕ
E^s	stable eigenspace
E^u	unstable eigenspace
E^c	center eigenspace
W^s_{loc}	local stable manifold
W^u_{loc}	local unstable manifold
W^c_{loc}	local center manifold
W^s	global stable manifold
W^u	global unstable manifold
W^c	global center manifold

References

1. Arnold, V.I., 1973, *Ordinary Differential Equations,* MIT Press: Cambridge, MA.
2. Gukenhiemer, J. and P. Holmes, 1990, *Nonlinear Oscillations, Dynamical Systems, and Bifurcations of Vector Fields,* Springer-Verlag, New York, NY.
3. Hale, J., 1980, *Ordinary Differential Equations,* Wiley: New York, NY.
4. Hale, J. and H. Kocak, 1991, *Dynamics and Bifurcations,* Springer-Verlag: New York.
5. Hirsch, M.W. and S. Smale, 1974, *Differential Equations, Dynamical Systems, and Linear Algebra,* Academic Press: New York, NY.
6. Li, T.Y. and J.A. Yorke, 1975, "Period Three Implies Chaos," American Mathematical Monthly 82 (10), 985-992.
7. Nitecki, Z., 1971, *Differentiable Dynamics,* MIT Press: Cambridge, MA.
8. Munkres, J.R., 1999, *Topology,* Second edition, Prentice Hall: NJ.
9. Royden, H.L., 1968, *Real Analysis,* Second edition, Macmillan: New York, NY.
10. Stokey, N.L. and R.E. Lucas Jr., 1989, *Recursive Methods in Economic Dynamics,* Harvard University Press: Cambridge, MA.

Index

Oded Galor is Professor of Economics at Brown University and a Fellow of the Department of Economics at the Hebrew University of Jerusalem. He is the Director of the Minerva Center for Economic Growth, a Co-Director of the *NBER* research group on Income Distribution and Macroeconomics, a Member of the Academic Council of *DEGIT*, a Member of the Scientific Committee of *CRISS*, and a Research Fellow of *CEPR*.

Oded Galor has served on numerous editorial boards. He is the editor of the principle journal in the field of economic growth, the *Journal of Economic Growth*, and he is the editor of the newly established quarterly, *Foundations and Trends in economic Growth*. In addition, he has been a member of the editorial board of *Economics and Human Biology*, and the *Journal of Economic Inequality*, an associate editor of the *Journal of Money, Credit, and Banking*, and *Macroeconomic Dynamics*, and a member of the advisory board of the *Journal of Economic Research*.

He has published extensively in the leading economics journals (*American Economic Review, Econometrica, Journal of Political Economy, Quarterly Journal of Economics, Review of Economic Studies*) and he has made seminal contributions to the fields of Inequality and Growth, Unified Growth Theory, and Evolutionary Growth Theory.

Made in the USA
Monee, IL
07 July 2026

56546133R00096